PORSCHE SPECIALS

PORSCHE SPECIALS

By Lothar Boschen and Jürgen Barth
Translated by Peter Wareham
Edited by Paul Frère and Michael Cotton

Patrick Stephens, Wellingborough

© 1984 Motorbuch-Verlag, Stuttgart
English language translation © 1986 Patrick Stephens Limited

All rights reserved. No part of this publication may be reproduced, stored in a retrieval system or transmitted, in any form or by any means, electronic, mechanical, photocopying, recording or otherwise, without prior permission in writing from the publishers.

First published in Germany by Motorbuch-Verlag, Stuttgart, under the title *Das Grosse Buch der Porsche-Sondertypen und -Konstruktionen*, von Lothar Boschen/Jürgen Barth

First published in Great Britain 1986

British Library Cataloguing in Publication Data

Boschen, Lothar
   Porsche specials: from 1931 to the present day.
   1. Porsche automobile
   I. Title.   II. Barth, Jürgen.   III. Frère Paul
   IV. Cotton, Michael, *1938–*
   V. Das grosse Buch der Porsche. *English*
   629.2′222   TL215.P75

ISBN 0-85059-802-8

Patrick Stephens Limited is part of the Thorsons Publishing Group.

Printed and bound in Great Britain.

Contents

Introduction 7
Special types and designs
Aero-engines: Types
  70/72/678/702/PFM 3200 *8*
American Roadster: Type 540 *15*
Amphibious car: Type 128/166 *18*
Apal: 356 replica *21*
Argentinian coupé: 356 power
  unit *22*
Armoured vehicles: Types
  100/101/205/714/806/814/1906,
  etc *23*
Auto Union, Grand Prix car:
  Type 22 *34*
Behra-Porsche: Formula 2 *37*
Berlin-Rome car: Type 114/
  Type 64 *40*
Bertone Roadster: 911 Cabriolet *42*
Beutler-Porsche: Type 356-003 *44*
Beutler Special: 356 base *44*
Brickyard: Indy racing car *48*
Buchmann: Types 911/928/924 *50*
Cisitalia: Type 360 *56*
Classic Motor: 356 replica *59*
Cooper-Porsche: 356 engine *60*
Cross-country vehicle:
  Type 82/Volkswagen *61*
Cross-country vehicle:
  Type 597/Porsche *63*
Denzel: VW base *67*
Diba GTC: 911 base *70*
Dick four-seater: 911 base *72*
Drews: 356 engine *74*
Durlite-Porsche: 550 engine *75*
Elva: four-/six-/eight-cylinder
  engines *76*
Emergency car: 924 base *78*
Envemo: 356 replica *79*
Evex: 911 base *80*
Fageol: twin-engine 356 *82*
Flat Profile: 911 Turbo base *83*
Frua (Hispano-Aleman): 914-6 *85*
Garretson: 914-6 base *86*
Glöckler: 356 base/engines *87*
Goertz: 914-6 base *91*
Gordini: 550 engine *93*
GRIPS: street-cleaning study *94*
Hess: 356 motor cycle *94*
Hirondelle (Swallow): 356
  engine/chassis *95*
Hovercraft (Bonner): 356
  engines *96*
Hydrobus (hydrofoil): 356
  engine *97*
Industrial engine: Type 616 *98*
KMW-Sport: six-cylinders
  +turbo *100*
Kremer Racing: 935 base *102*
Landing-craft engines: Types
  170/171/174/174A *104*
London-Sydney: 911 base *105*
Long-life car (FLA) *107*
Marine engine: Type 729 *111*
Mathé Universal: special single
  seater *111*
Mercedes-Benz land-speed record
  car: Type 80 *112*
Mercedes-Benz racing engine:
  Type 94 *116*
Midget: 911 engine *117*
Mini-model engines: four/twelve
  cylinders *118*
Moped engine: Type 655 *120*
Murene: 914 base *120*
Neumann: 356 aluminium
  roadster *120*
Nordstadt: 911/928 engine *121*
Ojjeh: 935 street *123*
ORBIT: fire engine study/
  Type 2567 *127*
Ostrad traction engine:
  Type 175 *130*
Pick-up: 914-6 base *131*
Pininfarina: 911 four-seater *132*
Police Porsche: Types
  356/911/912/914/924 *135*
Poll-Platje: 356 engine *138*
Pooper: Formula 2 *139*

Porsche: Type 52 sports car *140*
Porsche: Type 514 *142*
Porsche: Type 530 four-seater *144*
Porsche prototypes: Types
  534/555/728 *146*
Porsche: Type 695 *150*
Porsche: 911 four-seater *152*
Porsche: 914 eight-cylinder *153*
Porsche: 928 four-seater *155*
Porsche-Allgaier: tractors, Types
  110/111/113/425/535 *156*
Porsche-diesel: Types
  2086-2089 *161*
Research car: Type 995 *163*
Rossi: 917 base *164*
Sauter: 356 base *166*
SAVE: ambulance study
  Type 2539 *167*
Sbarro: 911 Turbo engine *170*
Silver Satin: 356 base *171*
Simpson-Chevy: 911 base *172*
Steel Special: 911 Coupé *173*
Storez: Carrera-GT base *174*

Studebaker: Type 542 *175*
TAG-Turbo: Formula 1 engine *178*
Tapiro: 914-6 base *181*
Through-flow water turbines: Types
  291/292/285 *183*
Townes: 911 base *184*
VW mid-engined saloon:
  Type 1966 *186*
Volkswagen (Beetle): Type 60 *188*
Wanderer: Type 7 *190*
Wheel-loader: 80/120/200 bhp *191*
Wind-power generators: Types
  135/136/137 *192*
World cycling record: 935 base *194*
World record cars: Types
  908/917-30 *196*
Zimmer: 910 base *198*
Weissach: the ideas factory *201*

Appendices
1: Various special types and
  designs *210*
2: Porsche type number index *217*

Introduction

From the outset, when Professor Ferdinand Porsche founded his technical design studio in Stuttgart, the firm of 'Porsche' was more than just a design office for the blueprint and production of new car prototypes.

The type number index of this sports car manufacturer provides an insight into the whole range of Porsche's activities, containing a great deal of interest outside the sphere of vehicle production (see *The Porsche Book* by Lothar Boschen and Jürgen Barth, Patrick Stephens 1983). This ranges from the Auto Union Grand Prix cars with revolutionary engineering techniques, military tanks, outboard motors, aircraft engines and tractors through to the Wanderer saloon car. Contract work for the German Federal Ministry for Research and Technology (BMFT) has resulted in studies for the sports car of the future, quiet street-cleaning machines, a new concept of emergency rescue vehicle, and a novel fire-fighting system.

No problem is too great for the Porsche research centre at Weissach to tackle; no solution too difficult. It is here that world-beating sports and racing cars have been produced, which in recent years have won countless championships.

But despite every effort on Porsche's part to provide customers with a wide product range, there were always enthusiastic amateurs, tuning specialists, engineers and racing drivers, with varying degrees of skill, who modified the original Porsche sports or racing car.

Private developments did not always go to plan, but this book attempts to do justice to these efforts, and to describe the less well-known designs from Porsche over the last fifty years, as well as the most interesting conversions by individuals all over the world.

The Porsche company has produced a wide variety of designs, over eighty in all, intended for use on land, sea and air. However, there had to be a limit to what we could reveal from Porsche's drawing boards, for two good reasons: hidden behind many type numbers are blueprints, test rigs and studies that would mean very little to readers outside a coterie of well-read engineers. Again, Porsche is bound to secrecy in respect of outside contracts, and even in cases where co-operation took place decades ago, some clients are not prepared to release information.

So, in order to avoid speculation, secret activities—as well as the less interesting projects—have been omitted. Some may still surface in the future.

Lothar Boschen

SPECIAL TYPES AND DESIGNS

Aero-engines: Types 70/72/678/702/PFM 3200

The name of Porsche has been associated with aeronautical engineering since 1908, many years before Professor Porsche founded his company. Ferdinand Porsche was the technical manager of Austro-Daimler in that year, when the company produced its first airship engine.

The first aero-engines from the Porsche design studio appeared in 1935/36. The first was a 32-cylinder, water-cooled radial engine with a swept volume of 17.7 litres (Type 70), and a sixteen-cylinder engine in 'V' formation, also water-cooled, came out at almost the same time with a capacity of 19.7 litres (Type 72); Professor Porsche was specially honoured for his aero-engine developments which came at a busy time, when the P-Wagen Auto Union racing car was in ascendancy and the KdF (Kraft durch Freude = strength through joy) car was under development.

Not all aero-engines were such giants, though. In 1943 a tailless aircraft took off with a 33 bhp Volkswagen flat-four engine and a pusher airscrew. The engine was mounted in the fuselage behind the cockpit, and drove the folding propeller by means of a 'V'-belt. An electric starter motor allowed the engine to be stopped and restarted during flight, and the maximum speed of this 'Porsche power-driven glider' was around 87 mph (140 km/h).

Twelve years passed before Porsche's engineers were again active in the aeronautical field. At the German Industrial Fair in 1959 the company exhibited a new range of power units at the Hanover-Langenhagen airport, and particular interest was shown in the Type 678/4 which had an initial power output of 75 bhp.

Two aircraft were demonstrated with Porsche power units: the RW3-Multoplane with the Porsche 678/4 engine, and the Elster from Pützer using the 678/1 engine; both were two-seater 'planes with a single engine.

After more than three years of test-bed running and flight trials the family of Porsche aero-engines had grown to four, all air-cooled, either with fan cooling or by propeller backwash cooling. All four were type-tested for international approval, and overhauls were scheduled after 600 flying hours.

The particular attributes of the Porsche aero-engines included high inclination tolerances, good starting under all climatic conditions, high temperature tolerance, versatility of mounting (due to the availability of different cooling systems), compact dimensions, and easy maintenance.

Derived fom Porsche car engines, the dimensions will be familiar to 356A owners: bore and stroke of 82.5 × 74 mm, with a capacity of 1,582 cc. The electrical system was 12V, and two sparking plugs were installed in each cylinder combustion chamber. Valves were operated by pushrods and rockers,

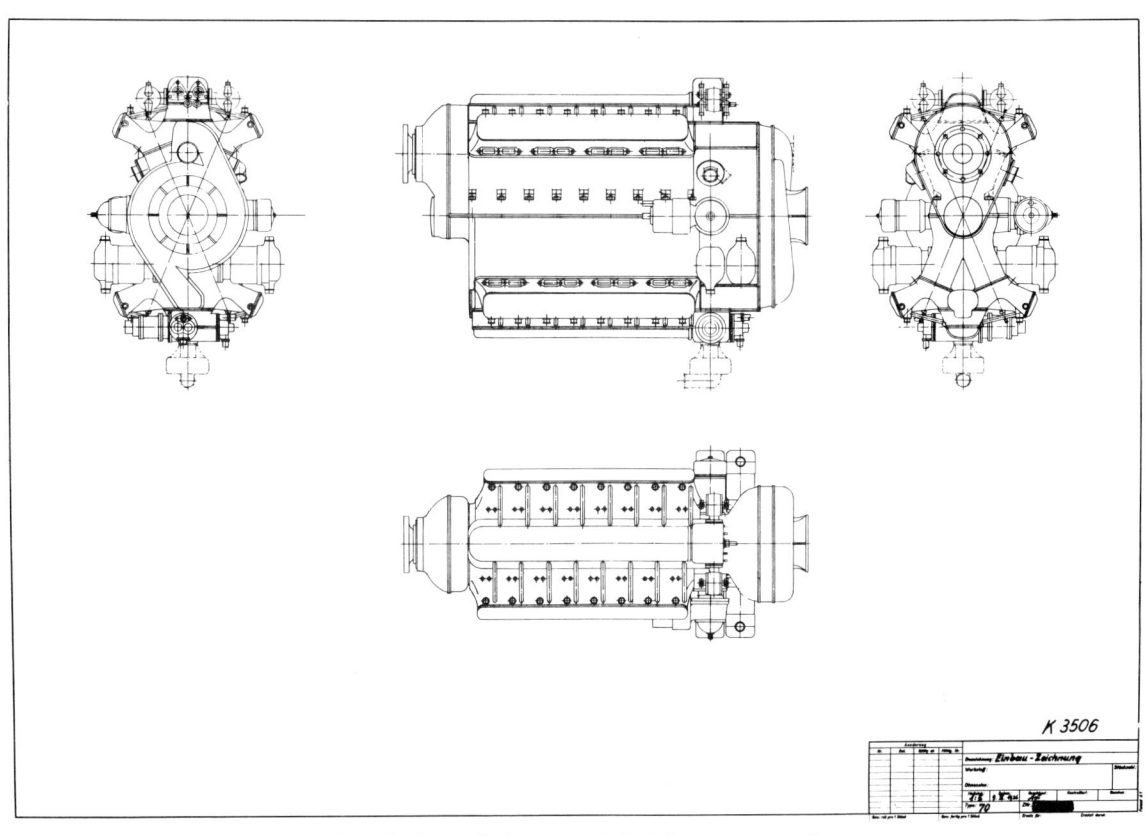

Type 70 aero-engine: a 32-cylinder radial motor of 17.7 litres swept volume.

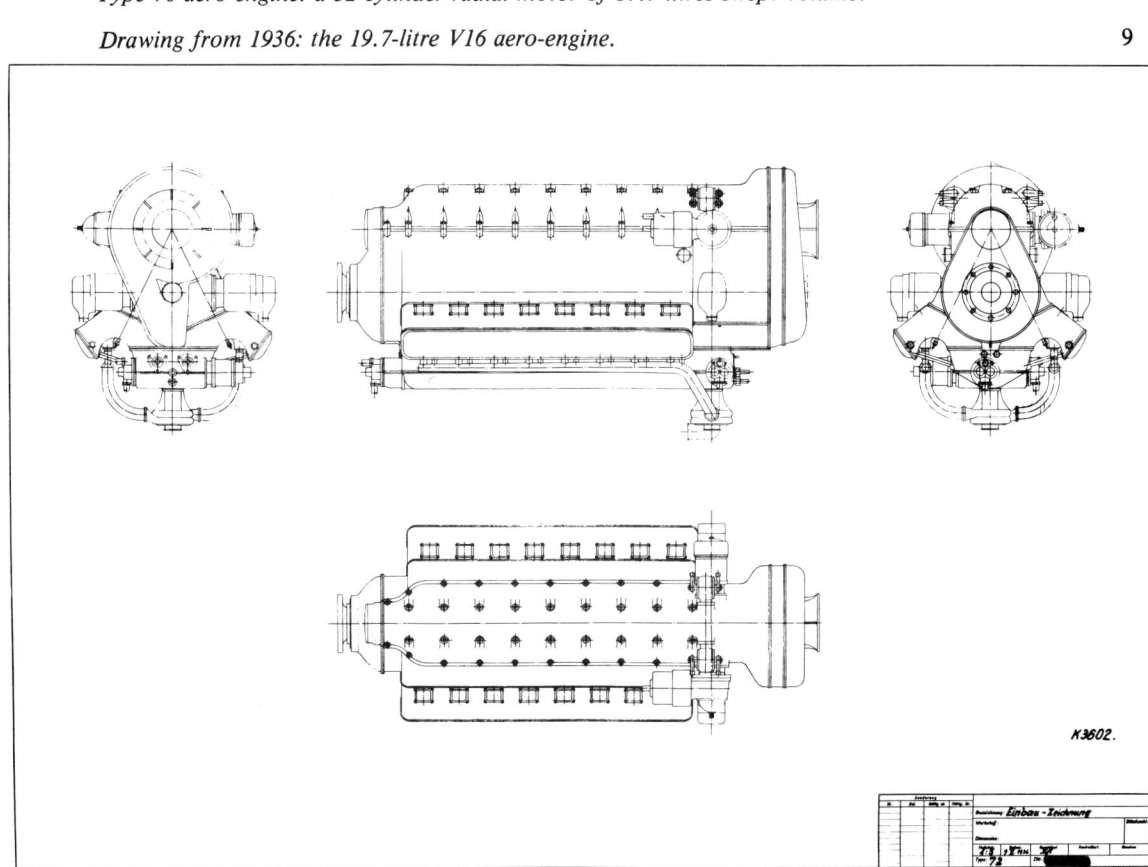

Drawing from 1936: the 19.7-litre V16 aero-engine.

Above *Introduced in 1959, the 678/1 aero-engine produced a continuous 55 bhp, weighed 233 lb (106 kg) and cost DM 4,940.*

Left *Type 678/3: continuous 50 bhp, air-cooled, with twin carburettors. The price was DM 4,115.*

and there was one overhead valve per cylinder on the inlet side, another on the exhaust. Low grade RON 80 aviation fuel was specified.

The Porsche aero-engine Type 678/1 had a starting power of 65 bhp at 4,500 rpm, and used about 4.6 gal (21 litres) of fuel per hour. Another version rated at 55 bhp at 4,200 rpm had a consumption of 3.73 gal (17 litres) per hour. Both were cooled by the air-stream, and featured single downdraught carburettors. Dry sump lubrication systems were installed, and the engines had a compression of 7.5:1. Optional ignition systems were twin batteries, or a dual magneto system. Without oil they weighed approximately 234 lb (106 kg), and were priced at DM 4,940.

The Type 678/3 had a starting power of 52 bhp at 3,200 rpm and a consumption of 3.62 gal (16.5 litres) per hour. Another version offered 50 bhp at 3,250 rpm and used 3.41 gal (15.5 litres) per hour. Again, cooling was by the air stream, but this version had normal oil sump lubrication, two special carburettors, HT magneto dual ignition and, without oil, weighed 187 lb (85 kg). The price was DM 4,115.

The Type 678/3A version had a 12V dynamo and an electric starter motor, which added DM 230 to the price.

The more powerful Type 678/4 had a starting power of 75 bhp at 4,600 rpm and a consumption of about 5.3 gal (24 litres) per hour. Another version was rated at 70 bhp at 4,500 rpm, and used 5.25 gal (22 litres) per hour. In this version cooling was by means of a radial fan driven by a dual 'V'-belt, and a flat tubular air/air oil cooler. Twin downdraught carburettors were used, other features including dry sump lubrication, a 12V starter motor, and dual batteries or dual magnetos. The compression ratio was 9:1, and the engine weighed some 247 lb (112 kg). It was priced at DM 5,860, and came on to the market in 1959, replacing the Type 678/0.

Porsche's first association with helicopters dates as far back as 1918. In the final stages of the First World War Professor Porsche (then, general manager of Austro-Daimler) built a helicopter for the Österreichische Flugzeugfabrik—the Austrian Aircraft Factory.

This helicopter never flew, in fact, but was used to carry out preliminary trials. Porsche's idea was that a helicopter must be powered by a particularly light engine, otherwise it would have disappointing performance. He therefore designed an electric motor producing 300 bhp but weighing only 550 lb (250 kg)—a quite astonishing design feat at that time.

A good forty years later the helicopter cropped up again, this one being the one-man machine from the American company, Gyrodyne, which was powered by Porsche's Type 702 aero-engine. It made its debut at the 1961 Paris Air Show, and straight away won a performance test in its class.

Twenty years further on the AD 500 airship made its successful maiden flight. The British company Aerospace Developments London (later, Airship Industries) had developed a new type of airship and a pair of Porsche 911 SC engines were chosen to power it. With a top speed of 72 mph (112 km/h) the AD 500 is certainly slower than any Porsche down below, but is faster, smaller and lighter than any previous airship design.

With a payload of 2.5 tons, and a range of 24 hours, the AD 500 shows significant advantages. At a height of only 525 ft (160 m) the craft is inaudible from the ground, for at a cruising speed of 48 mph (77 km/h) the 911 SC engines produce only 60 dB of noise. The AD 500 is therefore suitable for advertising work and coastguard duties, among other things.

The Type 678/4 aero-engine: continuous 70 bhp, radial fan, twin downdraught carburettors, priced at DM 5,860.

A one-man helicopter from Gyrodyne/USA powered by the Porsche Type 702 aero-engine.

The Airship Industries AD 500 airship is fitted with two Porsche 911 SC engines.

The outer skin of the AD 500 airship is made of reinforced polyester, coated with polyurethane, and is filled with the inert gas, helium. The gondola is also made of plastic, and can carry up to fourteen passengers.

The two Porsche 911 SC engines are virtually in standard road form, apart from modifications to the Bosch K-Jetronic fuel injection system to enable them to run properly at a maximum height of 33,000 ft (10,000 m).

Meanwhile development is continuing in the aero-engine sector, and, after two years of development work on the six-cylinder, air-cooled 911 SC engine, PFM 3200, flight tests began in September 1983, the full certificate of airworthiness being granted mid-1984.

Further modifications were needed to meet the stringent requirements of the Federal German Aviation Authority, which requires among other things that electrical systems must be duplicated. Twin spark-plug cylinder heads have to be fitted (like those on previous racing engines, such as the RSR), a second, separately operated dynamo, and a duplicated battery-coil ignition. The camshaft drive was changed from chain to spur gears and the pistons had to be modified, too. The output of the cooling fan was increased to ensure adequate cooling.

The power of the engine has actually been reduced, from 230 to 210 bhp, and under full-load conditions it uses about 9 gal (40 litres) of pump fuel per hour. Trials were carried out in a Cessna 182 aircraft.

Following modifications for aeronautical use, the six-cylinder engine provides 210 bhp.

Testing in a Cessna 182 aircraft showed that the Porsche engine uses 9 gal (40 litres) of 4-star petrol per hour under full load.

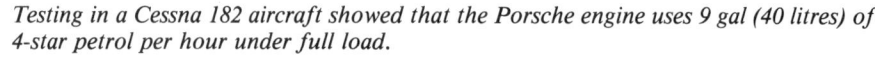

American Roadster: Type 540

The forerunner of the open Spyder, called the American Roadster and based on the 356, was a racing Porsche modified by Heinrich Sauter (see page 166). About a year after the Sauter Porsche appeared, early in 1951, Porsche KG built the 'American Roadster' exclusively for export.

The floorpan was the same as that of the Cabriolet, which was strengthened by adding metal to meet the anticipated higher stresses. In order to save weight, aluminium was used for the body and the interior was kept strictly functional. Without door panelling, and with detachable instead of wind-up windows, and sparse trim, this first 'racing version' of a Porsche 356 weighed about 1,575 lb (715 kg) and was fitted with the 1.5-litre engine developing 70 bhp. If an American Roadster was to be used for racing, more weight could be lost. There were, for instance, alternative windscreens, lighter seats, a leather strap to secure the bonnet, and the tonneau could be removed.

Between early 1952 and winter 1953 some twenty American Roadsters are supposed to have been built, with bodies supplied by the firm of Gläser in Ullersricht. In fact the actual number of cars made is not certain, since the transition from the American Roadster to the 540, with many modifications, took place very gradually.

A rarity: American Roadster, chassis 12.317, made in July 1952.

Twenty American Roadsters are thought to have been made between April 1952 and the end of 1953.

The bodywork came from the firm of Heuer at Weiden, north of Munich.

The divided windscreen could also be removed.

The bare sports cockpit, with a Porsche 356 dashboard.

The 1500 S engine developed 70 bhp at 5,000 rpm (this is engine no 40035).

Amphibious car: Type 128/166

In June 1940 Porsche's design studio received an order to extend the design of the extremely successful Kübelwagen (Type 82) to include an amphibious version. Very soon afterwards a design was produced showing a car capable of negotiating land and water, with a tub-shaped body on the platform of the KdF car (Type 60).

In contrast to the Kübelwagen, the amphibious version (still identified by the type number 128), was provided with four-wheel drive and an extra cross-country gear. While the detailed drawings were still being finalized, a quickly converted Kübelwagen underwent initial trials, and the first prototype did some laps of the Max-Eyth lake in Stuttgart in September 1940.

All together thirty prototypes were built, Porsche holding on to just a few for further trials. The Type 128 was given the standard Volkswagen power unit, and torsion bar sprung axles were retained at the front and rear. The only new feature was a ZF self-locking differential for each axle.

The Type 128 was driven by the 1,131 cc Volkswagen engine, developing 25 bhp. In addition to the four forward gears, it also had the specified cross-country gear driving all four wheels, selected when needed by means of a lever.

Only the rear wheels were driven when the car was going along normal roads, and the wheels continued to rotate when the vehicle was in water. The amphibious drive, however, was provided by a three-bladed screw attached to the rear, which needed to be swung down into position.

With five people on board, the amphibious car could move through water at 6 mph (10 km/h), with a good 30 in (80 cm) of draught, while on land it would reach 50 mph (80 km/h). The 128 easily met its design criteria, and in April 1941 the order was given for a subsequent model, Type 166. The first prototypes were undergoing trials four months later, and production versions started to roll off the lines in Wolfsburg early in 1942.

By the end of 1942, 511 amphibious Type 166 cars had left the factory and production built up rapidly, so that by the end of 1944 a total of 14,263 cars had been built.

Technically, the 166 was based on the 128, differing in some details. The later model was considerably more compact, but still provided room for four fully equipped soldiers. With virtually the same unladen weight, the 166 had a larger fuel tank extending the range from 275 to 325 miles (440 to 520 km). In overall dimensions it was 15¾ in (40 cm) shorter than the 128 and 4 in (10 cm) narrower, although the mechanical components were almost identical.

Porsche's amphibious car, Type 166, with the canopy lowered (April 1942).

The amphibious car based on the successful Kübelwagen, Type 82.

From the Type 128 prototype of 1940, the first '166' vehicles were produced in 1941. The 1.1-litre VW engine producing 25 bhp was mounted at the rear, and the model's cruising weight was 2,006 lb (910 kg).

Although shorter and narrower than the 128, the 166 still had room for four fully equipped soldiers.

Apal: 356 replica

The Automobiles Apal company in Belgium has been in business for many years, specializing in the production of glass-fibre replicas. Right from the start this company concentrated on Volkswagen and Porsche, mainly on the Type 356: Apal made a special body based on the 356 Super 90, which looked very similar to the famous Abarth Carrera.

The 356 Speedster Replica attracted more attention at the end of the '70s when the boom in older Porsches began. Hidden beneath the very faithfully reproduced bodywork are mechanical elements from the VW Beetle, such as engine and chassis. The Porsche Apal Speedster offered as a complete kit is assembled, sprayed and includes upholstery and electrical system, but lacks the chassis when priced at around DM 23,000.

The chassis is taken from a VW Beetle but shortened to Speedster dimensions, and matches the plastic body kit. Apal will also carry out this work on supplied second-hand frames, or even DIY frames assembled by a mechanic. The kit is designed for use with a 1.6-litre VW power unit developing at least 50 bhp.

A well-made replica of the Porsche 356 Speedster is offered by Apal, the Belgian company, and is based on the chassis and components of the VW Beetle.

Another view of the VW-based 356 replica as produced in Belgium by Apal.

Argentinian coupé: 356 power unit

During the 1950s a company in Argentina built its own sports coupé using a Porsche engine and transmission. Plastic was used for the construction of the two-door, two-seat coupé, and although the car, powered by a Porsche 1.5-litre, four-cylinder engine, aroused a good deal of interest, only a small number were made, for cost and manufacturing reasons.

Armoured vehicles: Types 100/101/205/714/806/ 814/1906, etc

From the early 1930s the Porsche design studio was involved with armoured vehicles and their variants, but so complicated is the subject that a specialist author had to be called upon to assist with the classification. In his book *German Military Vehicles and Tanks from the First World War to the Present Day*, Werner Oswald deals critically with the early forays by Porsche into the armoured vehicle business.

This is Oswald's analysis of the Porsche 'Leopard' tank, Type 100:

'In May 1941 Hitler demanded a heavy battle tank and, on 20 April 1942, Henschel and Porsche each presented a prototype. The Henschel model was very conventional, while Porsche provided his model with many technical innovations. It used, for example, two air-cooled diesel engines of his own design, each with V10 cylinders and 300 bhp. When it became clear that the development time for the engines alone would be too long, an air-cooled sixteen-cylinder diesel engine in X-configuration was selected for the tank. But, as with the V10, development of the sixteen-cylinder never started, and Porsche turned instead to a diesel-electric power unit.'

According to Oswald, 'Hitler's favourite tank designer' built interesting, but overly complicated designs which would be too difficult to produce. In the demonstration of the prototypes the Henschel also revealed some deficiencies, but these were of a more minor nature and could be eliminated by subsequent development.

'The Porsche prototype failed completely,' says Oswald. Although all the experts believed that the Henschel 'Tiger' should be developed, Hitler ordered preparation for quantity production of both tanks. Henschel delivered the first Tiger for service in August 1942 and manufactured a further 83 by the end of the year. The Porsche Tiger (Type 101, an improved version of the Type 100 Leopard), was to be built at the Steyr works, but the extremely complicated technology meant that only about ten were made . . . and these were never ready for the front line!

For this reason the existing 90 chassis were fitted instead with the Maybach carburation power units, rather than the still unfinished Porsche engines, and the Alkett company provided the appropriate superstructures. This hybrid tank destroyer was officially called 'Elephant', but in Oswald's words, 'this striking design blunder was nicknamed Ferdinand.'

In 1943 new orders were issued for the design of a completely new generation of tanks. With six basic models—E 5/10/25/50/75/100—ready to be copied, Porsche's studio did not participate. However, quite independently, Porsche was still developing his own tank ideas. These covered designs 245, 250, 255 and 205. Particularly interesting here is Type number 205, a mobile bunker weighing 197 tons (200 tonnes) and bearing the rather modest name 'Maus' (Mouse). Two prototypes, almost completed by the end of the war, were blown up shortly before the arrival of the Russians. From the parts which were found later, it was assumed that construction of about ten 'Mice' was planned.

About fifteen years later, the Porsche name cropped up again in tank engineering: in 1959 the German Federal Office for Defence Technology commissioned two firms, in competition, to design a new standard tank for the Bundeswehr and to supply two prototypes apiece.

The consortium of Porsche, Atlas-MaK, Luther and Jung was able to provide the more convincing design at the presentation, in 1962. The Porsche tank (Type 714), impressed through its shape and the ten-cylinder Daimler-Benz diesel engine, with an existing gun from England.

The decision went to the Type 714 and Krauss-Maffei was commissioned to produce it in volume. First, though, the 28 prototypes already manufactured (1960–62) had to be followed by fifty pre-production tanks which were built in 1962 and '63. In September 1965 Krauss-Maffei delivered the Leopard, as it was now officially called, in its final production form. A total of 2,300 Leopards were built in four sizes, A1 to A4, eventually costing DM 1.7 million apiece.

In 1970 a further contract was issued by the German Federal Ministry of Defence. This time Krauss-Maffei, Porsche and Wegmann were to develop an advanced tank, the 'Leopard 2' based on the first design. A V12 engine manufactured by Daimler-Benz, and developing 1,500 bhp, was to be used in Leopard 2, coupled to a Renk automatic gearbox. By 1975 Krauss-Maffei had built twenty prototypes for evaluation, and they looked to be the world's best battle tanks. Compared with Leopard 1 they had even more manoeuvrability, exemplary armoured protection and improved fire power. At the time of its introduction, in the autumn of 1979, the Leopard 2 cost DM 3 million. By 1986 the Bundeswehr should have more than 1,800 tanks of this type in service.

Since 1945 the Porsche project group for defence technology has taken a leading role in the following projects, through from concept stage to volume production: Leopard 1 and 2; armoured recovery, bridge-laying and pioneer vehicles; pioneer systems; self-propelled Howitzer 155-1; armoured cars; 'Luchs' (Lynx) armoured reconnaisance vehicle; 'Wiesel' (Weasel) self-propelled gun.

Leopard (Type 100)

At the beginning of 1940 a design was produced for a medium-heavy tank weighing 34½ tons (35 tonnes), with petrol-electric power and carrying a 7.5 cm gun. In order to save space inside the tank, new drive assemblies (toggle joints) with longitudinal torsion bars were provided, which were secured to the body on the outside. On either side of the tank were six rubber-tyred double track rollers, the upper part of the chain track being supported by a pair of rollers.

At the rear were two air-cooled, ten-cylinder engines and associated generators, developed by Porsche. Transmission was provided by two electric motors located one behind the other at the front (centre) of the tank, and these provided the track drive by means of sun wheels and planetary gears. The slipping clutches built into the electric motors protected the drive system against damage.

The trials provided valuable information on electric control and steering systems, and on the behaviour of air-cooled engines in a tank construction. During the initial trials demands were made for more powerful weaponry and armour plating, and this resulted in the early abandonment of the trials. A design rethink led to the 'Tiger' (Type 101).

Dimensions
Track width 8 ft 6 in (2,600 mm); *Chain width* 2 ft (600 mm); *Ground clearance* 1 ft 7 in (490 mm); *Width* 10 ft 6 in (3,200 mm); *Firing level* 7 ft 3 in (2,220 mm); *Track ground contact length* 10 ft 7 in (3,225 mm); *Track roller diameter* 2 ft (600 mm); *Length* 21 ft 8 in (6,600 mm); *Height* 9 ft 11 in (3,030 mm); *Weight* 34½ tons (35 tonnes).

Porsche's 1940 Leopard tank (Type 100) was technically complicated.

The Type 100 failed in its trials, but production was nevertheless ordered in 1941. Tests at the Steyr works. Professor Porsche is on the far right in the turret.

Tiger (Type 101)

The Tiger tank was a direct successor to the Leopard (Type 100), incorporating a number of improvements. Based on Type 101 there was a further development, Type 102, which featured a hydraulic drive instead of electric drive. The hydraulic unit was supplied by the Voith company and, although fifty units were supplied, only one was fitted. The Type 102 ceased production before the end of the war. The specification of the Tiger is as follows:

Engines
2 × ten-cylinder V construction: *Bore* 115 mm; *Stroke* 145 mm; *Cubic capacity* 15 litres; *Compression* 5.9:1; *Revs* 2,400 rpm; *Maximum power* 2 × 320 bhp = 640 bhp; *Starter motor* Bosch AL/SED with auxiliary motor type 141; *Drive* Petrol-electric, electric motor at the rear; *Steering* Continuous electric; *Maximum speed* 22 mph (35 km/h); *Wheels* 31¼ in (794 mm) diameter steel (rubber damping); *Track width* 8 ft 8 in (2,640 mm); *All-in width (incl chains)* 13 ft 8 in (4,175 mm); *Chain width* 2 ft 1 in (640 mm); *Overall length* 30 ft 8 in/22 ft (9,340/6,700 mm); *Overall width* 10 ft 4 in (3,140 mm); *Overall height* 9 ft 2 in (2,800 mm); *Permissible total weight* 54–56 tons (57–59 tonnes); *Fuel capacity* 115 gal (520 litres); *Armament* 1 × 8.8 cm KWK 36 gun L/56 (70), 2 × MG 34 machine guns; *Gradient* 30°; *Vertical obstacle* 2 ft 7 in (780 mm); *Fording* 3 ft 3 in (1,000 mm); *Trench* 8 ft 10 in (2,640 mm).

The Porsche 'Tiger' tank (Type 101) was the improved version of the Leopard.

Only about ten Tiger battle tanks were built: two of them are seen here.

The Tiger tank at Steyr's Nibelungen works, before completion.

The tank destroyer with the Leopard chassis, nicknamed 'Ferdinand'.

Maus (Type 205)

This enormous tank, the largest ever made, reached only the prototype stage: two were made, but neither got into action. An interesting feature, shown in the drawing, is the special construction of a railway train which was to transport the Maus overland. The specification reveals just how colossal this project was:

Overall length (gun to the front) 33 ft 1 in (10.085 m); *Overall length (gun to the rear)* 41 ft 6 in (12.659 m); *Overall length without gun* 29 ft 8 in (9.034 m); *Overall width* 12 ft (3.67 m); *Overall height (without aerial)* 11 ft 11 in (3.63 m); *Ground clearance* 1 ft 10 in (0.57 m); *Fording* 6 ft 6 in (2 m); *Maximum speed* 12½ mph (20 km/h); *Permissible continuous speed* 11 mph (18 km/h); *Fuel consumption (road)* 0.2 mpg (1,400 litres/100 km); *Fuel consumption (moderately difficult terrain)* 0.07 mpg (3,800 litres/100 km); *Range (road)* 115 miles (186 km); *Range (moderately difficult terrain)* 42 miles (68 km) with drop tank, 26 miles (42 km) without.

Engine

Type Daimler-Benz 509; *Cylinder arrangement* suspended V; *Cubic capacity* 44.5 litres; *Continuous power (boosted)* 1,080 bhp at 2,300 rpm; *Cooling fan power* 120 bhp.

Only two Type 205 'Maus' battle tanks, weighing 185 tons (188 tonnes), were manufactured.

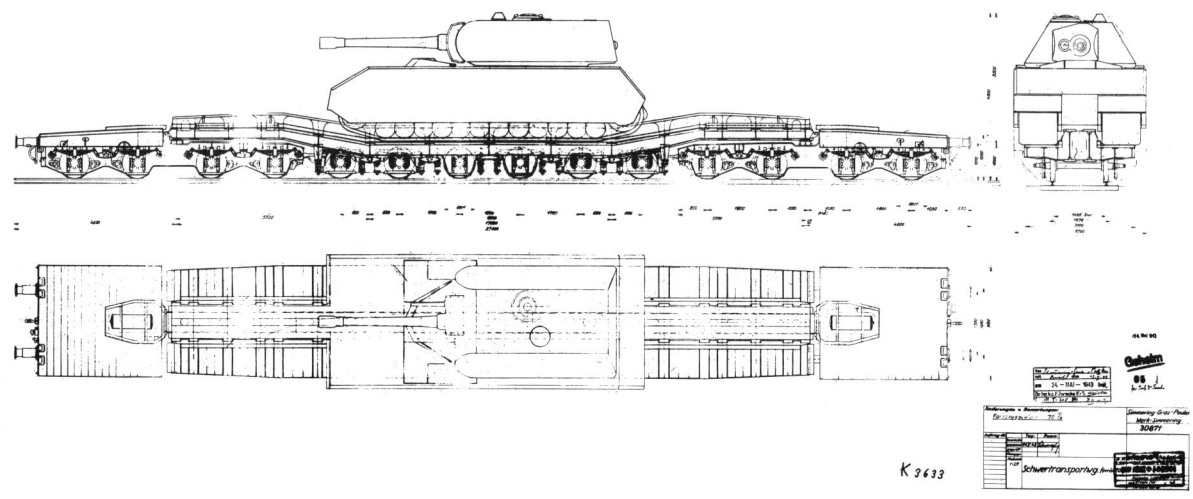

The means of transport for the gigantic 'Maus' was a special railcar.

Drive
Electric drive (twin generators and two electric motors) each 400kW, 800V; *Gear ratio between electric motor and chain drive wheel* 1:25.44; *Gear ratio between electric motor and chain drive in hill gear* 1:48.68; *Number of track rollers per side* 24; *Suspension system* Porsche toggle joint with equalization action; *Track roller system* Steel rings, rubber damping.

Undercarriage
Chain support length without ground pressure 19 ft 4 in (5.9 m); *Chain width* 3 ft 7 in (1.1 m); *Ground pressure at 8 in (20 cm) sinking width* 18.63 lb/sq in (1.31 kg/cm^2); *Wheel pressure, average* 7,900 lb (3,580 kg); *Brake* One self-adjusting disc brake per chain in lay shaft; *Oil pressure* 60 atmospheres overpressure; two foot pedals as stationary brake, with two hand levers and cable; *Steering* Electric with two steering levers and governor.

Armament and ammunition
1 KWK 12.8 cm L55 gun; 1 KWK 7.8 cm L36 gun; 1 MG 151/20 machine gun in turret; 1 MP sub-machine gun in turret; 1 smoke mortar; *Ammunition* Rounds and shells divided as follows—68 × 12.8 cm rounds (25 in the turret, 43 in the hull), 200 × 7.5 cm rounds (125 in the turret, 75 in the hull).

Special equipment in the vehicle
1 auxiliary petrol engine with 48V/5.5 kW generator for battery charging, engine cowling heating and auxiliary drive; 1 alert system; 1 fire extinguishing system; 3 drainage pumps; 1 acetylene cold-start system with Dissous acetylene gas cylinder.

'Standard' armoured recovery vehicle

A special armoured vehicle related to the Leopard was developed by Porsche in co-operation with MAK and Jung. With a 20-ton crane, dozer blade, recovery winch and extensive range of equipment, the armoured recovery vehicle is designed for all eventualities.

It has a simple and rugged construction following the principle, already established by Professor Porsche, of designing extra strength for all the heavily stressed parts, and making everything else as light as possible. The armoured recovery vehicle is easy to drive and operate, but demands versatility and technical skill of its four-man crew.

Its main specification is: length, 22 ft 9 in (6.94 m); width, 10 ft 8 in (3.25 m); height, 8 ft (2.4 m); engine cubic capacity, 37.4 litres; power 830 bhp at 2,200 rpm. Weight (according to equipment) between 39 and 41 tons (40–42 tonnes). Gradient, 60 per cent, fording 3 ft 11 in (1.2 m), deep fording 7 ft 5 in (2.25 m) and underwater fording down to 13 ft (4 m). Maximum speed 40 mph (65 km/h).

The Daimler-Benz-powered Leopard tank (Type 714) went into production in 1963.

'Biber' armoured bridge-layer

The extensive Leopard family also includes the armoured bridge-layer developed in co-operation with Porsche. Together with their two partners Porsche had carried out extensive studies concerning type and applications in order to find uniform systems for bridges and bridge-laying. The design proposes a bridge with a triangular cross-section consisting of ramp and middle sections which can be coupled together.

A length of up to 92 ft (28 m) can be transported by the bridge-laying vehicle. For extensions to 138 ft (42 m) or more, appropriate sections must be added. The vehicle carries out the coupling and laying of bridges without outside help. These are conveyed on rollers, and supported by a hydraulically driven bridge-laying girder which also helps to lay them.

The 'Biber' armoured bridge-layer (Type 807) is one of the new Leopard generation.

'Wiesel' self-propelled gun

For the mobilization of weapons systems for the infantry, a vehicle was needed which offered adequate protection for the crew. Also, the entire system had to be capable of being transported by air.

Within the many variations of the Wiesel (Weasel) family, the power unit, chassis and seating are identical modules. The drive system is located at the left front, next to the driver. A mass-produced petrol engine (Audi EA 381, type 75) engine with 100 bhp is used. The five-speed gearbox is from the Porsche 928 car series.

A converter is placed between the engine and the gearbox, allowing clutchless drive so that the engine cannot stall. The reverse gear is operated by means of hydraulic steering brakes, allowing continuous steering radii from 8 ft 6 in (2.6 m) to infinity (on the longitudinal axis). The cooling system is designed for a temperature range from −30°C to +44°C (−22°F to +110°F). The exhaust system complies with German road traffic requirements.

The 'Wiesel' self-propelled gun has a 100 bhp Audi petrol engine and a Porsche 928 gearbox.

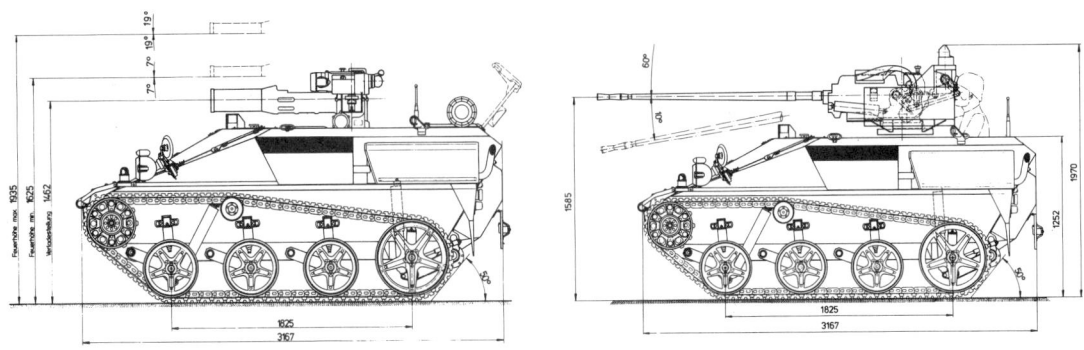

A service brake acts on the two end drive shafts using ventilated brake discs. The front-drive vehicle is supported by eight rollers (four on each side), with torsion bar suspension. The Wiesel is 10 ft 6 in (3.2 m) long, 5 ft 11½ in (1.82 m) wide and 6 ft (1.83 m) high, and the heaviest version weighs 6,725 lb (3,050 kg).

Leopard 2 (Type 1906)

The latest generation of Leopard battle tanks is 25 ft 4 in (7.73 m) long, 11 ft 7 in (3.54 m) wide and 8 ft 2 in (2.49 m) high, and has a fighting weight of just over 49 tons (50 tonnes). The Daimler-Benz engine is a twelve-cylinder unit with exhaust-driven turbocharger, 39.8 litres in capacity, and develops 1,500 bhp at 2,600 rpm. The gearbox alone weighs 4,625 lb (2,100 kg), having four forward and two reverse gears.

An additional power generator is provided in the form of a two-cylinder multi-fuel engine, developing 22 bhp as a diesel, 20 bhp with petrol or 9 kW generator power. Leopard 2's maximum speed is 42 mph (68 km/h), the gradient capability around 60 per cent, and the tank can move under water at a depth of 18 ft (5.5 m).

Leopard 2 and 'Leo 1': Porsche's Weissach centre did most of the development work on this very successful design.

Auto Union, Grand Prix car: Type 22

When, in October 1932, the former International Sports Commission decided on a new formula for Grand Prix racing to apply from 1934, Professor Ferdinand Porsche was one of the first to start work at his design studio. The new formula left the engine size completely free, but imposed a *maximum* dry weight of 750 kg (1,653 lb) without tyres.

A new company was founded for this project under the name of 'Hochleistungs-Fahrzeugbau GmbH' (High-Performance Vehicle Construction Co Ltd), and by March 1933 the drawings were produced for the Grand Prix car. It was to have a V16 engine of 4.36 litres capacity, and a Roots supercharger.

Up to this time Porsche had worked without a sponsor, and now he looked around for firms who could finance or take over his project. Through his association with Wanderer, for whom he had developed a six-cylinder saloon at about the same time, Porsche established contacts with the new Auto Union combine, of which Wanderer was a member.

When it became known that Daimler-Benz had also built a racing car for the new formula, Auto Union took up the challenge and bought the 'P-Wagen' design from Porsche. A racing department was set up at the Horch factory in Zwickau, and a 130-strong team began work on this unusual design. The engine was mounted behind the driver, though ahead of the rear axle, and below this ran the driveshaft to the rear-mounted gearbox. The fuel tank was located between the engine and the driving seat, at first with a 37-gal (170-litre) capacity, later to carry 46 gal (210 litres). The front of the car carried the oil tank and water radiator, and an oil cooler was soon added.

The drivers had first to get used to the unconventional driving position just behind the front axle, and in fact the concept was not taken up again by Porsche until the 1970s, in the 908-03.

On 12 January 1934 the Auto Union Grand Prix car was revealed for the first time to a group of prominent people. Hans Stuck drove on the Avus circuit without any problems at over 125 mph (200 km/h), and a couple of months later he made a record attempt on the same track. With an extended, more aerodynamic tail on the P-Wagen, Stuck set up a new one-hour world record of 217.11 km/h (134.9 mph), reaching a top speed of 300 km/h (185 mph).

The original P-Wagen, later classified as 'Type A', had a 4.3-litre engine which, with a compression ratio of 7:1, developed 295 bhp at 4,500 rpm. The sixteen-cylinder engine was not in fact the most powerful in Grand Prix racing but, with some justification, Professor Porsche saw the advantage of enormous torque, 54 mkg at only 2,700 rpm.

Porsche's design was sensational in some respects, and a combination of these revolutionary innovations and some more familiar features in contemporary design made the car unique. A feature which was advanced for its time (and therefore soon abandoned!) was the routing of the water lines from the front radiator to the engine, and back again, through the long chassis tubes.

Since the fuel blend at that time was unrestricted, each company had its own secret mixture. We know now that Auto Union, for instance, ran their engines on a mixture of 60 per cent alcohol, 20 per cent benzol, 10 per cent diethyl ether, 8 per cent petrol, 1.5 per cent tolulene and nitrobenzene, and 0.5 per cent castor oil.

The Auto Union single-seater's first success was not on a race track but in the mountains. The 'Bergmeister' Hans Stuck won the hill-climbs at Felsberg (Saar) and Kesselberg on two successive Sundays. The first major victory, though, came when Stuck won the German Grand Prix on 15 July 1934 in the Auto Union (coded Porsche Type 22) from the Porsche design studio.

By that autumn work had already begun in the Horch racing department on Type B, the successor to Type A. The engine capacity was increased to 4,951 cc, and the compression ratio raised to 8.95:1. The engine now gave 375 bhp at 4,800 rpm, roughly 30 per cent more than Type A's. Although the wheelbase was lengthened by 4.3 in (110 mm), Type B was actually shorter overall and slightly lower. The entire cooling system was changed radically and the rear axle was made lighter by dispensing with the heavy transverse leaf springs and replacing them with torsion bars.

Hans Stuck tested a special aerodynamically improved version, the 'Lucca' record car, on the Florence to Lucca autostrada on 15 February 1935 and covered the flying mile at 199.01 mph (320.267 km/h)!

During that year the Type E scored seven major race victories, but plans were already being made for 1936, when the Type C was unveiled. Again the engine capacity was increased, this time to 6,005 cc, the compression ratio rose to 9.2:1 and power to 520 bhp at 5,000 rpm. The torque had reached a sensational 87 mkg—and this in 1936!

There was no doubt that the Auto Union Type C Grand Prix car was the pace-setter in 1936, with the highest performance and the best preparation. It scored nine major wins, seven of them with Bernd Rosemeyer driving—a marvellous record. Nevertheless the 'battle' commenced, Mercedes presenting the Type W 125 for the 1937 season with 5,660 cc and over 600 bhp. By contrast, Professor Porsche altered only a few details on the Type C for the anticipated confrontation with Mercedes.

The 1937 season marked the end of the '750 kg formula' which had begun in 1934, and of the dozen most important races, Mercedes won seven with the W 125, Auto Union only five, reversing the 1936 score-sheet. And the end of this formula also saw the end of Professor Porsche's involvement with the Auto Union racing department, responsibility for subsequent types between 1938 and 1940 being assumed by Dr Robert Eberan von Eberhorst.

The Type C was also the basis for another record-breaker which Bernd Rosemeyer drove at over 250 mph (400 km/h), on 25 October 1937. The single-seater was almost fully enclosed, and its engine was bored out to 6.33 litres so that it developed 550 bhp. In other respects, though, it was the same as the Grand Prix engine.

When Auto Union learned that Mercedes had hopes of bettering this achievement, the 1937 record car was further developed in great haste. The engine was now bored to 6.5 litres and the bodywork was further streamlined, giving the Type C a theoretical maximum of some 283 mph (456 km/h)!

Early in the morning of 28 January, Caracciola drove his Mercedes at an average speed of 268.86 mph (432.692 km/h). Rosemeyer countered this in the Auto Union at 266.6 mph (429 km/h) on a warm-up run but then, as he was tackling the record in earnest, a gust of wind caught the car which left the road and crashed heavily. Rosemeyer sustained a broken neck and was killed.

Above *The Type C of 1936: initially it had a 6-litre capacity, giving 520 bhp at 5,000 rpm.*

Below left *The Grand Prix car with mid-positioned engine—the sixteen-cylinder power unit sits in front of the rear axle.*

Below right *A Roots supercharger feeds the engine with the air/fuel mixture under pressure.*

Behra-Porsche: Formula 2

The French Grand Prix driver and long-distance specialist, Jean Behra, joined up with Porsche in the '50s. At the time he had driven for BRM (1958) and Ferrari (1959), among others, in Formula 1 races, but in 1958 Behra had driven the so-called 'central-seater' Porsche to victory in the Rheims Formula 2 race.

The 'central-seater', a modified Porsche RSK, definitely offered certain advantages on the fast circuits, but it could not stay with its rivals on the slow tracks. Behra therefore decided—at the same time as the Porsche factory—to build a proper Formula 2 car. He needed an experienced engineer for the job, of course, and soon hired the Italian, Valerio Colotti, who had been involved in the development of the famous 250F Maserati and now owned a small firm in Modena specializing in the manufacture of racing gearboxes.

The concept of the Behra-Porsche was soon established: it would have a suitable aluminium body on a lightweight, compact tubular frame with an RSK engine and other Porsche mechanical components.

This straightforward project progressed quickly, and early in 1959 the Modena firm of Neri e Bonaccini completed the single-seater. In the meantime, though, Behra had signed a contract with Ferrari, so the Behra-Porsche had to wait a few more weeks for its debut.

This coincided with another debut—that of the Porsche factory's Formula 2 car, in the Monaco Grand Prix. The Italian driver Maria Teresa de Filippis was at the wheel of the Behra-Porsche, but she failed to qualify. The works Porsche F2 was not much luckier, though Count Berghe von Trips qualified safely among the well-subscribed Formula 1 grid (Formula 2 cars were eligible

Behra-Porsche: ordered by the Grand Prix driver Jean Behra in 1958. After a few detours it ended up in the States.

only if they made sufficiently good times). An accident early in the race forced von Trips' retirement, ending Porsche's hopes.

The high-point of the 1959 season was, again, the Formula 2 race at Rheims, the scene of Behra's triumph the year before. Still Behra was unable to drive the Porsche due to contractual difficulties, so Hans Herrmann got his chance and rose to the occasion, swapping the lead constantly with Joakim Bonnier in the works Porsche, Stirling Moss in the Cooper-Borgward and von Trips in the older 'central-seater'. Herrmann was narrowly defeated by Moss, but had the satisfaction of beating Bonnier in the works Porsche!

Midway through the season Behra fell out with the Italian Ferrari team, and at last had the chance of driving his own Formula 2 car in a race. For the German Grand Prix, held that year at Avus, he entered his Formula 2 car, and his Porsche RSK Spyder in the supporting sports car race on the previous day.

Tragically, Behra was killed in the sports car event. His Behra-Porsche did not start in the main race, of course, and the Porsche factory also withdrew as a mark of respect for the well-liked Frenchman.

After Behra's death the Camoradi team took over the single-seater and entered it for a number of races, among them the Argentinian and Italian Grands Prix. The Behra-Porsche was well-placed in both these events, but technical development was progressing rapidly and the Behra-Porsche ended up in America following the break-up of the Camoradi team.

The car was bought in the States by Vic Meinhardt, who ran it on an amateur basis. Then, when the veteran no longer stood a chance even in club racing, it was retired and thoroughly restored.

The cockpit of the Behra Formula 2 car, designed by Colotti in Italy.

The Formula 2, well used in competitions, shown after restoration.

Berlin-Rome car: Type 114/Type 64

While working on the 'Volkswagen' (People's Car) in Professor Porsche's design studio, the engine designers were already experimenting with higher-performance versions of the VW engine, for a sports model. But, at the end of the 1930s, those in power had no time for such frivolities!

Nevertheless, Porsche played around with a few ideas, at least on the drawing board. Between 10 and 16 September 1938, blueprints were prepared for an aerodynamic VW-based sports car; they bore the designations 114, 114 K1 and 114 K2, differing only in detail. They were, however, identical in the positioning of the drive unit: a modified VW engine in front of the rear axle—a classic mid-engined design.

It had a wheelbase of 110½ in (2,750 mm), front and rear track widths of 53 in (1,350 mm) and a turning circle of 34 ft 6 in (10.5 m). The K1 was designed to carry three people, the third sitting crosswise behind the front occupants; but the 'emergency seat' was omitted from the K2 design.

All these plans remained on file until a planned motor sport event quite unexpectedly presented a use for them. A long-distance race from Berlin to Rome was planned for the autumn of 1939 and, since the first 375 miles (600 km) of this long-distance race were over the new autobahns, the designers anticipated certain advantages, particularly from streamlined bodies.

As a few VWs were to take part in the event, Professor Porsche was given the go-ahead to build a special version of the KdF 'Beetle' (Type 60), while his colleagues continued without a break on the development of the Type 114. The first drawing for the Type 64, dated 9 January 1939, was based exactly on the Type 114, although with a slightly modified chassis.

By combining the aerodynamic body and the '60 K10' chassis (KdF car) Porsche produced three vehicles which were given the internal designation 'Type 64'. These had 985 cc engines producing 32 bhp at 3,800 rpm, and a maximum speed of about 80 mph (130 km/h). Torsion bars were used for the front and rear suspensions, and four mechanical internal-expanding brakes slowed the cars which weighed 1,875 lb (850 kg). The wheelbase measured 94½ in (2,400 mm), while the track was 51 in (1,290 mm) at the front and 49 in (1,250 mm) at the rear.

By the time testing began, the VW engines had been uprated to 40 bhp, and the maximum speed rose to 87 mph (140 km/h). The driver's seat was shifted almost to the middle of the car and the co-driver had to make do with a lightweight seat moulding, which was removable, mounted slightly behind the driver.

The outbreak of the Second World War led to the cancellation of the Berlin to Rome road race, so the three Type 64 sports cars never turned a wheel in competition. One was destroyed in a bombing raid, another was driven by the Americans until it was reduced to scrap. The third, though, came through the war unscathed and was acquired by the Austrian, Otto Mathé. Although disabled by an arm amputation, Mathé drove the car in post-war races up to 1951—and very successfully, too.

Wooden model (Type 114) for the later Berlin-Rome car (Type 64).

Of the three prototypes with 40 bhp engines, only one survived.

Bertone Roadster: 911 Cabriolet

Johnny von Neumann, manager of Volkswagen-Porsche-Audi-Pacific in the USA, was envious of the importers of other European sports cars. Even with their top models Ferrari and Maserati offered cabriolet versions, which Porsche did not. The 356 series had come to an end and with it the Speedster, Cabriolet, Convertible, Roadster and hardtop variants; and the 911 was available only as a coupé.

Neumann made contact with Bertone, the Italian designer of custom bodies, in 1966 and sent him a half-finished 911. From this, Bertone was expected to conjure up a roadster which would appeal to the American market, at a cost not higher than $8,000.

The Italian threw himself enthusiastically into the project and a few months later, at the Geneva Salon of 1967, introduced the Porsche 911-Bertone. This unique car took pride of place between the Jaguar FT and the Lamborghini Miura, and indeed attracted the most attention, its straight, clean lines representing the classical Italian sports car tradition.

Bertone even succeeded in shaping the bodywork so that the rear-engine configuration of the 911 was disguised. Only a row of air vents on either side, at the back, hinted at the air-cooled six-cylinder engine. The softly rounded nose, no longer even resembling a 911, hid the retractable headlamps behind a fine grille. Parking lights and indicators were incorporated in the wrap-round bumpers. The rear bodywork fell elegantly to a vertical cut-off, and there was a neat solution to the housing of lamps and indicators. Low down, under the bumper, was a horizontal grille to provide ventilation for the engine.

The interior was radically different from the standard 911's. The three main instruments were placed vertically—speedometer, rev-counter and oil pressure/temperature gauge one above the other—in the centre console. The open bodywork met right behind the front seats, a small flap concealing the soft-top and/or an extra seat.

Despite very favourable reaction, Neumann's project was doomed. The hierarchy at Stuttgart was ready to bring out the Porsche 911 Targa, and nothing was to detract from it. So the Bertone Roadster remained a one-off, although during 1967 it was given the more powerful 911S engine, and was later resprayed and re-upholstered.

Ordered by Neumann, the US Porsche dealer, in 1966, the Bertone 911 Roadster was exhibited at Geneva in 1967. The rear bodywork is almost completely enclosed, concealing the six-cylinder engine.

Beutler-Porsche: Type 356-003

The firm of Beutler Brothers in Dürrenast, near Thun in Switzerland, received an order in the autumn of 1948 to build six cabriolets for Porsche. Unlike the Reutter cabriolets built from 1950 onwards in Stuttgart, these had aluminium bodies, fully retractable soft-tops and the bonnet inscriptions were in individual lettering.

The first Beutler cabriolet appeared at the Geneva Salon early in 1949, while the last one was delivered in August to the Porsche agent, Blank, in Zurich. The cars were not difficult to sell at SFr15,500 apiece, but in any case Switzerland was a good market for Porsche cars: of the initial series of fifty cars made in Gmünd, in Austria, no fewer than twelve were sold in Switzerland.

The second Beutler Porsche, a Bordeaux red two-seater with a lighter-hue bench seat (chassis number 003) appeared in June 1949. It won an award in a *concours d'élégance* and was sold to Egypt, but reappeared in Switzerland in 1952. Then, in 1960, this same Beutler cabriolet was acquired by H.P. Wyssmann, from Berne. The original 1,086 cc engine had been replaced by a 1.3-litre power unit, and a synchromesh VW gearbox and telescopic shock absorbers had been fitted. The odometer showed 116,261 km (72,242 miles), of which 85,088 km (52,871 miles) had been covered with the replacement engine.

Its low weight of 1,460 lb (662 kg) helped '003' to achieve impressive performance figures. Just the same, high speeds took their toll of the narrow 16-in diameter wheels. At an average speed of 62 mph (100 km/h), fuel consumption figures of around 37.5 mpg (7.5 litres/100 km) could be obtained.

Beutler Special: 356 base

The chaos of war had forced the Porsche company to move to Gmünd in Austria—safer during hostilities but creating problems at the end of the '40s when the new sports car, Type 356, was being prepared. Porsche enlisted the aid of the Swiss firm Beutler to build the first of fifty cars, and not until the Porsche design studio moved to Stuttgart in 1950 was most of the vehicle production brought in-house.

Porsche did not forget that Beutler had been a good partner in difficult times. The coachbuilder therefore received an order in 1957 for some experimental versions of the 356, which by then had become extremely popular.

The Beutler, with VW chassis, Porsche brakes and a 1.6-litre Porsche engine (1957). It no longer resembled a Porsche sports car, from any viewpoint.

The first 'Beutler Special' made its debut in 1957 at the Geneva Salon. It was technically unimaginative, without any great inspiration, and reminded viewers of the Borgward Isabella. Parts of the Beutler Special were taken from the VW Beetle and Karmann Ghia, and modified. For the two-door saloon Beutler took the floorpan from the Beetle, while the engine, chassis and brakes came from Porsche. The front section was identical to the Porsche 356's, the sides and rear being modified by Beutler.

Beutler offered this four-seat Special for sale, to order, in 1957 and established a production level of about ten cars a year. As time went on the firm made a number of detail bodywork modifications, and built cabriolet versions.

A second series which went into production in 1960 was more convincing. This Beutler Special, offered as a coupé or cabriolet, bore more of a resemblance to the Aston Martin or Sunbeam Alpine which were popular at the time. Beutler also experimented with a four-seater 356, which still had the unmistakable nose of the original 356 on its extended aluminium body, while the roof was shaped like a hardtop and extended right to the back. These Beutler versions generally added around DM 20,000 to the price of a 356.

Despite this considerable extra cost, Beutler built and sold about sixty coupés or cabriolets based on the Porsche 356. Apart from the high price, the Beutler Specials had the disadvantage of a considerable increase in structural weight, and the 60 bhp Porsche engine was unable to give them a decent performance.

The second Beutler series (from 1960) was reminiscent of an English sports car. The coupé and cabriolet versions were built only to order.

In 1960 Beutler also experimented with a four-seater Porsche 356. The nose remained original, but the aluminium body was drastically modified at the rear. The hardtop roof conversion was priced at DM 20,000.

Brickyard: Indy racing car

After a quarter of a century of activity in international motor sport the Porsche factory had, up to the mid-'70s, taken part in most categories: Formula 1, Formula 2, prototypes, Groups 5 and 7, Can-Am, Grand Touring and Touring classes as well as rallying. For many reasons, though, it had never managed to get to the famous Indianapolis 500.

However, there were some private drivers who nurtured the belief that they had a chance of success with Porsche engines in this popular American race. Albert H. Stein, a midget racing car champion from Oakland, California, tried a spectacular combination in 1966—a twin-engined Porsche, specially built for Indy qualifying. Stein was not the first to attempt the 'Brickyard' race with a twin-engine single-seater, though. Lou Fageol (see page 82), a fellow country-man, had already made the attempt in 1946 with two Offenhauser engines, and later built a car with two Porsche units, although he did not run that at Indianapolis. Albert H. Stein was therefore the first to compete at Indy with Porsche engines.

Stein had already begun planning a twin-engined racing car in 1963, but his ideas came to fruition only in 1966 when a friend from Germany offered him three 901 (911) engines at a price which undercut the cost of expensive Offenhauser or Ford power units. The Californian reckoned that two six-cylinder Porsche engines must be at least as good as the home-grown engines!

Joe Huffaker, the racing car constructor, was commissioned to build a suitable chassis, and delivered a tubular steel frame ready for the next stage of the operation. A Lancia racing gearbox, Girling disc brakes and magnesium wheel rims were fitted, and Stein fabricated an aluminium body.

The whole project was certainly very complicated. The front engine was positioned with the gearbox facing the driver, while the rear engine had its gearbox facing the back, with each power unit driving the appropriate set of wheels. The gearboxes had to be synchronized with a complex linkage. The two power units were controlled to operate largely in synchronization, but the front one could be shut down if necessary.

The 'Brickyard-Porsche' arrived at Indianapolis in May 1966 and the experienced driver Bill Cheesbourg drove it during qualification. Despite several attempts to put the twelve-cylinder 'Porsche' on the 33-car grid, however, he was unsuccessful.

The fastest lap for the car was logged at just 149 mph (240 km/h)—10 mph too slow. All further attempts failed, so Stein sadly packed up and drove back to California.

Stein then attempted to sell his creation to industry for the nominal price of one dollar, in the hope of raising further finance. No offers came forth, and nothing more has been heard of this 'Brickyard-Porsche'.

Above *Twin-engined Porsche—failed in the Indianapolis qualifying (1966). A speed of 149 mph (240 km/h) was achieved in qualifying: too slow by 10 mph.*

Right *Complicated: rear and front engines (Type 901) with the driver between.*

Buchmann: Types 911/928/924

Rainer Buchmann, manager of the 'b + b Auto-Exklusiv-Service' company in Frankfurt, started out in the mid-1970s. At the time the demand for unusual cars was at its peak and there were plenty of ideas around, although not all of them were practical. Buchmann and his team thought they had the answer though, as the Porsche range seemed predestined for the purpose. To start out, they prepared a 911 Targa with a successful, strikingly striped coachwork, while the interior fittings could be adapted to suit any taste.

The big breakthrough came with a model that was regarded as an unusual hybrid, for at the beginning of 1978 Buchmann and his best body specialist took a 3.3-litre 911 Turbo and stripped it right down. The bare bodywork of the Carrera Targa was equipped with Turbo parts and adapted to the different dimensions.

The front-end treatment was like that of a Porsche 928, and a large oil cooler was fitted under the Buchmann-developed rear spoiler. In the tail, the Turbo engine was boosted to 1.1 bar, and gave 370 bhp—some performance for a Targa! Appropriately wide rims and tyres were, of course, part of the package.

While the looks and performance were already spectacular, 'b + b' really went to town on the interior fittings: a television set, CB radio, stereo system, special seats, even a fridge were installed!

Including a full kit of gold-plated knobs, flashing lights and controls, this 911-Turbo-Targa-Carrera-928 was to cost 'only' DM 215,000. The confection was sold to a not impecunious collector in the Antilles.

Buchman immediately became the centre of attention, with Porsche and Mercedes drivers his preferred customers. Visual improvements on the 911, 924 and 928 were his speciality, and at the Frankfurt Show in 1979 he displayed the Porsche 928S Targa. This involved a modification to the bodywork above the waistline, while in contrast to the massive rump of the 928, the Buchmann variation used a short 'notchback' boot with a vertical rear window.

From certain aspects this gave the 928S Targa an even more elegant line, while more drastic changes in the conversion included treatment to the impact-absorbing nose section. The new front gave the 928 a more masculine and striking appearance—that, at least, was the opinion of its builder.

The removal of the roof, which called for bodywork changes and chassis bracing, involved a lot of work and expense. Tubular steel reinforcement was needed in the engine compartment, front wheel arches, door sills and under the rear seats. The top part of the windscreen frame was also reinforced with additional plates. But the inclusion of a longitudinal centre strut, between the top of the windscreen and the rear arch (the so-called Targa centre strut), had no load-bearing function . . . it served only as a home for the very expensive stereo system and the optional car telephone!

The revised front-end on the Buchmann 928 Targa was made of polyester, as well as the removable Targa roof which was stored in the boot. All the other elements, such as the roll-over hoop, boot lid and wing top sections at the rear, were hand-worked from sheet metal. All mechanical parts remained as standard.

With a delivery time of about three months, this conversion was priced at DM 35,000.

In 1979 Rainer Buchmann of b+b Auto-Exclusiv-Service, Frankfurt am Main, presented his Targa version of the Porsche 928 at the Frankfurt International Motor Show.

Conversion work above the waistline required expensive measures to brace the remaining bodywork. The conversion costs about DM 35,000.

Buchmann achieved his big breakthrough with a Porsche 911 Turbo, on to which a Carrera Targa body was mounted. The front treatment was similar to the 928, and the engine delivered a good 370 bhp at 1.1 bar boost. The interior was lavishly fitted with gold and leather accoutrements.

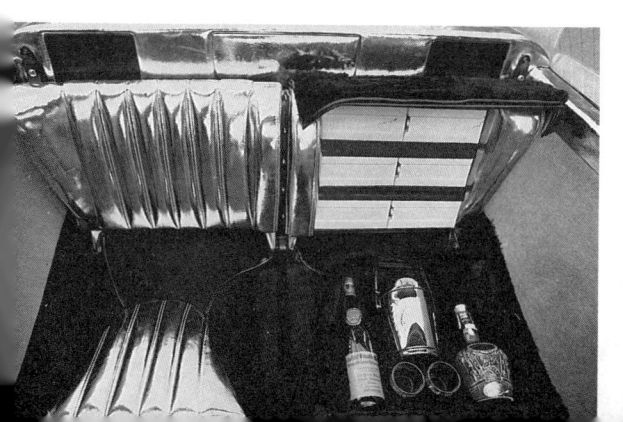

The first Buchmann project in the mid-1970s: an eye-catching striped design for the Porsche 911 Targa, with 'tea tray' tail spoiler.

The rather unimaginative bodyline of the Porsche 924 was enlivened by b+b with a new, colourful finish and wider rims and tyres.

Cisitalia: Type 360

In the immediate post-war years, before it was possible to mass-produce cars, Ferry Porsche received an interesting order. Piero Dusio, managing director of the Cisitalia works in Turin, wanted to enter racing in a big way with a real Grand Prix car. Dusio already made 1.1-litre single-seaters and sports cars in Italy, and they were running successfully.

Dusio intended to spare no expense on his way to winning Grand Prix races: the car was to be a 1.5-litre single-seater, with a supercharger and all the technical innovations then available.

Porsche's team went to work straight away, and by 1949 had produced the first and only 'Cisitalia' Grand Prix car. Even by contemporary standards it was technically sensational when delivered to Dusio, but after a series of unfortunate mishaps his firm went bankrupt, and the Italian found that he was no longer in a position to race the car.

At the time of these financial difficulties Dusio was in Argentina, where he had just set up the Autoar company—the first-ever independent Argentinian car factory. Autoar manufactured its own models, also building the NSU Prinz under licence. After a while, however, Dusio returned from Argentina and paid off all his creditors.

The 'Cisitalia', along with spare parts, was shipped to Argentina without this extremely expensive single-seater ever having been raced. It was put in a corner of the Autoar works with a dust-cover over it, and remained there until about December 1951 when, more or less by accident, a friend of Dusio discovered the treasure. Now began the long, laborious process of persuading a large number of people to bring the Cisitalia back to life.

A professor, who was a specialist in combustion engines at the university in Buenos Aires, took over all the documents and decided first of all to run tests with a single-cylinder experimental engine. A whole year passed while this research was carried out, and then came the 'Temporada 1953', a big racing spectacular which took place every year in January. It was here that the Cisitalia was to race for the first time, and Clemar Bucci, an Argentinian, was engaged to drive it.

The Cisitalia actually had its first outing only a few days before the race—on the roadway next to the Autoar factory. It was beset by problems, starting with the spark plugs and not ending with the tyres. Two days before the race the Argentinian mechanics and assistants had come to terms with the worst problems, and Bucci drove it for the first time.

He returned to the pits straight away complaining of a defective gearbox, although it is likely that he had not mastered the gear selection which, in motor cycle fashion, was in a straight line. Then Felice Bonetto took over for one lap in the Cisitalia. A former factory driver for Alfa Romeo, Maserati and Lancia, Bonetto found himself followed by a dense cloud of smoke, and it was discovered that the venting of the crankcase was not working properly, and oil was dripping on to the exhaust. At this point the car with withdrawn from the race.

In 1949 Porsche started to build the 'Cisitalia' Grand Prix car ordered by the Italian, Piero Dusio: twelve cylinders, 1.5-litre capacity, 450 bhp with Roots supercharger and 150 octane fuel, 187 mph (300 km/h), four-wheel drive.

Back at base the engine was removed from the Cisitalia and mounted on a dynamometer. Using an alcohol-petrol mixture the engine showed 385 bhp at 9,000 rpm, but disenchantment soon followed: the mixture was too weak and two pistons burned out, so it was back to square one.

As luck would have it, new pistons were found among the spare parts, but the mechanics were not trained to work on this complex engine and it became impossible to keep to a time schedule. Finally, in order to achieve *something*, it was announced that a record attempt would be made with the Cisitalia on 18 June 1953, and, although it was cold and windy, the Cisitalia appeared on time.

Clemar Bucci seemed to be very pleased with the steering and general performance after a test run, but again criticized the lack of engine power. The plugs were warmed up, the twelve-cylinder engine restarted, and the car reached a good 162 mph (260 km/h) in top gear at 8,500 rpm. But when the attempt was made to run in the opposite direction—necessary to set a record—a piston failed again due to weak mixture settings and the average of the two runs was only 145.5 mph (234 km/h), too little for the supercharged 1.5-litre engine, which needed to average at least 185 mph (300 km/h).

Autoar then abandoned all further plans for the car and offered it for sale . . . for just $4,000! It was bought by an amateur who took it away and rebuilt it as a dragster—using a Ford V8 engine. He did not have enough money even for this limited project, though, and the Cisitalia went back under a dust-sheet.

It was next seen in 1960 when the Porsche works team tracked it down while staying in Argentina. The Cisitalia was brought back to Stuttgart and refurbished for the Porsche museum, where it can be seen today.

Some technical data on the Cisitalia: capacity 1,493 cc; twelve-cylinder horizontally opposed engine, water-cooled, with four overhead camshafts driven by horizontal shafts; Roots supercharger; twin Weber horizontal carburettors; compression ratio 9.2:1; power 450 bhp at 10,500 rpm. It has a tubular spaceframe, with front wheel suspension on twin trailing links and torsion bars; at the rear, transverse wishbones with longitudinal ball-jointed arms and torsion bars.

The five-speed gearbox is behind the engine, and there is optional four-wheel drive via a universal joint propshaft. Two fuel tanks are located one on each side of the cockpit, filled with a special mixture of about 150 octane (present-day premium fuel is around 98 octane). The four-plate clutch runs in oil.

The weight of the complete engine is a mere 340 lb (155 kg). The size of the front tyres is 7.00-19 and the rear tyres 5.50-18 or 6.00-17. Wheelbase measures 100 in (2,600 mm); front and rear track 51 in (1,300 mm); maximum speed about 185 mph (300 km/h).

Classic Motor: 356 replica

The Classic Motor Carriages company based in Miami, Florida, includes two Porsche types in its extensive range of replicas: the 356A Speedster and a sports version of the same. The 200 employees of this American company produce some 300 replicas per month, either as a kit or assembled.

The kit for the 356A Speedster is offered by Classic Motors for $6,000 including bodywork and all necessary parts, together with most of the interior fittings. A good 90 per cent of this car is supposed to correspond exactly with the original. The customer needs only a VW Beetle chassis, to be shortened as specified, a Volkswagen engine, and a lot of time for assembly work!

Those wanting to go 'over the top' can order the Classic Speedster C, a racing version of the 356A replica. This differs from the normal Speedster in having all unnecessary decoration deleted, wider wings and tyres, a roll-over hoop, and power as required. The cost is a further $400.

Classic Motors in the USA offers the Speedster replica on a VW base.

Cooper-Porsche: 356 engine

Frequently to be seen in South Africa in the mid-1950s was a Cooper chassis with a Porsche 1500 S engine. With this car, Ian Frazer-Jones recorded a number of wins in important races. Of the original Cooper, only the nose remained unmodified.

Cooper chassis with Porsche 1500 S engine, raced in South Africa.

Cross-country vehicle: Type 82/Volkswagen

Other than by the 356—the ancestor of all Porsche sports cars—world-wide reputations have been established by the cross-country vehicle, Type 82, and the amphibian, Type 166.

The Kübelwagen (bucket car), as it was first called, appeared in design stage in May 1938. At that time it bore the type number 62, being based mainly on the KdF* car, which was Type 60. Trials with the KdF (later to become the Volkswagen Beetle) by the Wehrmacht had included tests of the Type 60 with the bodywork removed.

The resultant Type 62 was the forerunner of the cross-country vehicle, and early in 1939 the first working prototype was demonstrated. Further improvements led to the Type 82 design, which went into production at Wolfsburg in 1940 and was sold to the Wehrmacht initially for 2,782 Reichsmarks per car. In the next five years VW built 52,000 vehicles of this type, and they were used in countless forms on battlefields all over the world.

Until March 1943 the four-cylinder opposed engine had a capacity of 985 cc and developed 23.5 bhp. The capacity was then increased to 1,130 cc and the power went up to 25 bhp, although the compression remained at 5.8:1. Air cooling was provided by a fan, and carburation was by means of a Solex downdraught carburettor.

The single-plate Fichtel & Sachs clutch engaged four forward gears, and one reverse. At the rear, the swing axle was divided and equipped with a self-locking ZF differential. The front suspension was by twin trailing links and torsion bars; single-acting hydraulic shock absorbers damped the front suspension, double-acting shock absorbers the rear. Both the foot brake and the hand brake acted mechanically on all four wheels, which were disc type with 3.00 D/16 rims fitted with cross-country 5.25-16 tyres.

The Type 82 could cope with gradients of between 40 and 45 per cent, according to ground conditions, and had a maximum range of up to 280 miles (450 km), its consumption being 33.3 mpg (8.5 litres per 100 km).

The vehicle's vital statistics included a length of 147 in (3,740 mm), a wheelbase of 94½ in (2,400 mm), a width of 63 in (1,600 mm) and a height of 65 in (1,650 mm), the operational tare weight being 1,470 lb (668 kg). The lowest continuous speed that could be maintained was 2 mph, the maximum 50 mph (80 km/h).

---

*KdF = Kraft durch Freude (the Nazi slogan, 'Strength through joy').

Right and below *The Kübelwagen (bucket car) Type 72 was a Porsche development of the Type 60, dating from 1938. Production began in 1940 with a price to the Army of DM 2,782.*

Right *The Kübelraupe (bucket dozer) Type 155/1 was used by the Wehrmacht in Africa and Russia. It appeared in 1942.*

Cross-country vehicle: Type 597/Porsche

With all their wartime experience in developing the 'Kübelwagen' cross-country car and amphibious vehicles, Porsche felt they had a head start when the new German Army, the Bundeswehr, was formed in the mid-'50s with a requirement for cross-country vehicles. In a very short space of time—by the end of 1954—Porsche's Type 597 cross-country car was ready.

The chosen power unit was derived from the 1.6-litre 356 sports car, and developed 50 bhp, with a maximum torque figure of 10.7 mkg at just 2,400 rpm. An additional cross-country gear was incorporated in the four-speed synchromesh gearbox, making a total of five. Naturally the Porsche 597 was equipped with four-wheel drive, with a facility to disconnect the front drive for road use.

The steel platform chassis was self-supporting, and the all-independent

Porsche's Jagdwagen (hunting car) Type 597, as it appeared in prototype form early in 1950.

suspension system again relied on Porsche's torsion bar springing. Double-acting telescopic shock absorbers were fitted at the front and rear, and the foot-brake system was hydraulic to all four wheels, alloy drums being specified. The handbrake operated mechanically on the rear wheels.

The Type 597 used 6.00-16 tyres on 5.00 F-16 rims. The fuel tank had a capacity of 13 gal (60 litres) and even in the most extreme conditions the vehicles used no more than about 23.5 mpg (12 litres/100 km), reducing to better than 28.3 mpg (10 litres/100 km) in normal use. The unladen weight was registered as 2,403 lb (1,090 kg) and the vehicle was 142 in (3,600 mm) in length, 61½ in (1,560 mm) wide, and 62 in (1,580 mm) high with the canopy roof raised. Its lowest speed was 1.75 mph (2.8 km/h), top speed 62 mph (100 km/h), and it would climb a 65 per cent gradient in good conditions.

An unusual feature of the Porsche 597 was its ability to float, though without propulsion, so that it could be towed over water or attached to a transport convoy.

The prototypes were brilliantly successful in their trials with the Army, but, perhaps for political and financial reasons, the 597 was not selected; instead DKW obtained the contract with their 'Munga', powered by a two-stroke engine.

Left *The vehicle had selectable four-wheel drive for use on heavy ground.*

Below *In the back, the 1.6-litre air-cooled Porsche engine developing 50 bhp.*

Above *The Jagdwagen Type 597, with the NATO designation 'PKW 0.25 t GL', seen in 1955.*

Left *A special feature of the 597 was its ability to float, without propulsion.*

Below *The basic version of the 597 passed the toughest tests brilliantly in 1955.*

The 597 had a self-supporting steel chassis/body. The vehicle weighed 2,400 lb (1,090 kg) and had a top speed of 62 mph (100 km/h).

Despite good engineering, the Porsche 597 was not 'politically' acceptable.

The civilian version which appeared in 1958, with optional sidescreens.

Denzel: VW base

The heading 'VW base' indicates that this 1950s sports car does not necessarily belong in the category of Porsche projects. Nevertheless, although the Denzel was based mainly on VW parts, it was a product of its time and comparable with a Glöckler-Spyder or even a Porsche 550.

Wolfgang Denzel, the Austrian engineer and garage owner, had driven in rallies before the war, and after the hostilities he was keen to get back into competitions. In 1948 Denzel, with his co-driver Hubert Stroinigg, built a four-seater sports car—the Denzel—from old VW parts.

Its chassis was modified from a wartime VW cross-country car and power came from a 1.1-litre air-cooled, flat-four engine boasting about 25 bhp. The bodywork, due to a shortage of other materials, was made of wood! Since the wood was heavy, though, the first attempt was not promising, so the next version was given a plastic pontoon body. The Denzel now weighed 1,325 lb (600 kg) and proved competitive, enabling Denzel and Stroinigg to win the Austrian Alpine Rally. This success immediately resulted in enthusiastic enquiries about further examples.

Denzel was in business, and started manufacturing two- and four-seater cars in a Viennese suburb. The first step was to increase the puny output of the VW engine by increasing the bore, thus raising the output to 38 bhp. Porsche had just introduced his 356 on to the market, having modified the VW engine to produce 40 bhp.

In 1951 Denzel moved away from the backbone platform chassis of the VW Beetle, preferring a chassis with welded tubular steel sections. This was rugged, cheap, and was even lighter. The engine, transmission and suspension, though modified, still came from Volkswagen.

The following year Denzel replaced the plastic bodywork with an even more elegant, hand-made aluminium body, and the Denzel was built in this form until production ended in 1959. Only the engine output was steadily increased, so that eventually there were two power units: a normal road version with 1.3-litre capacity and 52 bhp, and a sports version with almost identical capacity but 64 bhp. On these cars only the engine block and fan housing were original VW Beetle parts; everything else was specially made by Denzel.

Almost at the end of the Denzel era the capacity was raised to 1.5 litres and, again, two versions were offered with 75 or 85 bhp. In almost twelve years of manufacture Denzel turned out around 350 models in roadster, cabriolet and hardtop coupé form.

As early as 1948 Wolfgang Denzel had already completed his first sports car. The pontoon bodies—aluminium on tubular steel—were built from 1952 to 1959.

Two 1.3-litre engines on a VW base were offered, giving either 52 or 64 bhp. About 350 Denzels were made, ending up with 1.5-litre engines and 75 or 85 bhp.

Diba GTC: 911 base

The platform of the Porsche 911 was used for a masterpiece, dubbed the 'Diba GTC', built at the beginning of the '70s by a Swiss bodywork specialist.

The two-door coupé had an 0.88 mm gauge all-steel body upon the Porsche 911 chassis with original mechanical and electrical components. The 2-litre six-cylinder engine produced 148 bhp at 6,100 rpm, and the 7 in wide special magnesium rims were equipped with 185 HR 70×15 Dunlop tyres.

The wheelbase was 87 in (2,210 mm); the front and rear tracks 52¾ in (1,340 mm) and 53¼ in (1,350 mm) respectively. Ground clearance was 6 in (150 mm), the length 170 in (4,320 mm), while the Diba was 70 in (1,780 mm) wide and 44½ in (1,130 mm) high.

Without fuel it weighed just about one ton (1,020 kg), had a total permissible laden weight of 3,086 lb (1,400 kg), and the unique GTC had a top speed of 134 mph (215 km/h). Consumption, according to the constructor, was between 15 and 16 litres per 100 km (18.8 and 17.7 mpg).

Diba GTC, built in the early '70s by a Swiss bodywork specialist.

An all-steel coupé body was mounted on the 911 chassis.

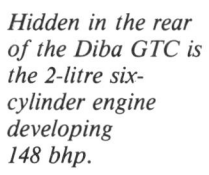

Hidden in the rear of the Diba GTC is the 2-litre six-cylinder engine developing 148 bhp.

The interior was taken mainly from the 911.

Dick four-seater: 911 base

William Dick Junior, co-owner of a Porsche agency in America's south-west, wanted to give his wife a very special gift: a Porsche 911 with four doors. Such a machine has never been offered by Porsche, of course, so Dick made his own arrangements, talking first to a company that undertook conversions of this type. Their proposal bore little resemblance to a 911, as it turned out. Dick therefore continued his search and found Troutman and Barnes in Culver City, California, and with the help of industrial designer Charles Pelley produced plans for a four-door 911 that did not alter the original silhouette too drastically.

The car that was made available to Troutman & Barnes was a 1967 911S with a Sportomatic gearbox. Extending the chassis by 21 in (533 mm) was of course the main problem, and accounted for the majority of the cost. With the help of a sheet-metal worker the base assembly, from the windscreen A-pillar back to the new rear window, was redesigned, thus providing the basis for incorporating four doors, all of which had to be the same size. To make this possible the two rear 911 doors were hinged at the back rather than at the front, so that the door handles and locks were located at the centre on each side.

The interior was upgraded with leather and walnut panelling, but the original dashboard was retained. The unusual four-seater 911 was presented to Mrs Dick at the end of 1967.

Years later, and after several changes of ownership, considerable modifications were made. These included replacing the original 2-litre engine with a 2.7-litre six-cylinder, and the Sportomatic was replaced by a five-speed manual box. The new owner, Dan Burns, provided the 911 four-seater with 7 × 15 in rims and powerful air-conditioning from an Audi saloon, together with a new finish.

Opposite page *The four-door Porsche 911, ordered by William Dick (USA) in 1967. With additional rear-hinged doors, the body was 21 in longer than the basic 911.*

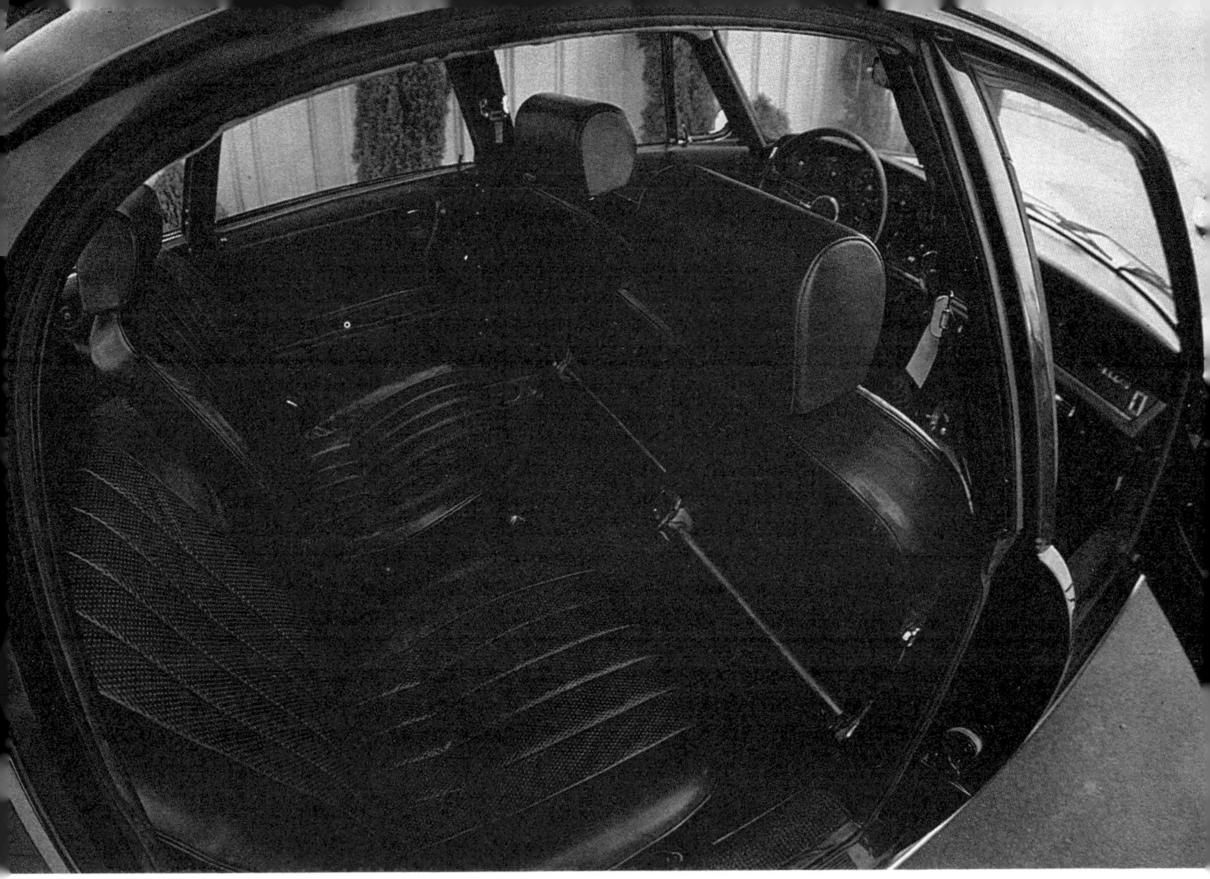

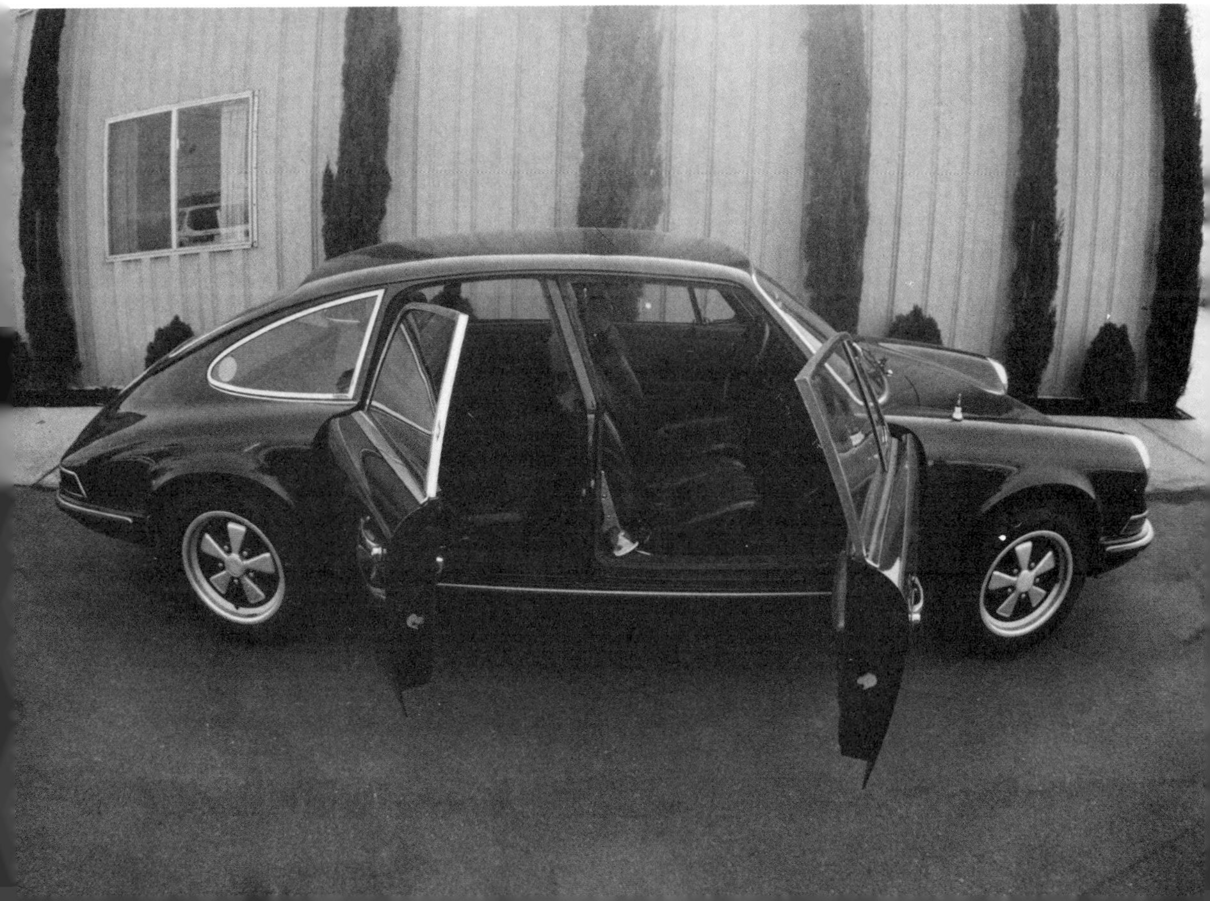

Drews: 356 engine

Strictly speaking, this sports car from the Wuppertal company of Drews does not belong in a book on Porsche specials but, at the beginning of the '50s, Drews did produce at least one coupé with a Porsche power unit.

The three brothers, Gerhard, Erwin and Werner Drews, had already built a car in 1947 good enough for Emil Vorster to win a national championship sport title with it. For their sports cabriolet produced shortly afterwards they, like many others, took a complete VW base assembly including the mechanical systems. New bodywork was hand-made and welded to the chassis, and the high labour content resulted in the price of a 'Drews' rising to about DM 10,000—a small fortune at that time!

Greater efficiency was achieved by producing the cars in batches of ten, greatly reducing costs. Since the general design, including the interior and fittings, was taken over from Volkswagen, the car's performance was also almost identical, although a 'Drews' was slightly lighter than a Beetle.

Among approximately 150 Drews cabriolets built before 1951, there was one coupé known to have been fitted with a Porsche 356A engine.

Below and above right *In 1947 the Drews brothers began building their own VW-based sports cars. About 150 were produced between 1947 and 1951, including one with a Porsche engine.*

Durlite-Porsche: 550 engine

The Durlite Mark III came about as the result of a shortage of cash. In August 1959 Bob Webb had damaged his Porsche Spyder in a race and, for financial reasons, decided that he was not in a position to buy a new one. However with the help of a couple of friends he was able to undertake the construction of his own sports-racing car, in which the four-cam Porsche engine was installed.

A tubular frame chassis of chrome-molybdenum-steel was copied from the original, and the bodywork was made of aluminium. A plastic-coated aluminium fuel tank was placed next to the driving seat. The engine was taken out to 1,600 cc so that it went exactly into the class E category; using lightweight Mahle pistons, different carburettors and other modifications, the engine, which had a compression ratio of 9.8:1, developed between 140 and 165 bhp.

The reconstruction and improvements to the Porsche Spyder RS cost a good $2,500—much less than a new Spyder would have cost—and resulted in a highly competitive car.

Elva: four-/six-/eight-cylinder engines

Elva, the small but successful firm in Essex, England, built excellent sports and racing cars in the 1960s and these had a good reputation, particularly in the 2-litre racing category. Some racing drivers in the States thought the car good enough to ship over, and fit with Porsche racing engines.

The racing department at Porsche agreed to co-operate and further developed the four-cylinder engine for fitment in the Elva chassis. At just 1.7 litres capacity and with a compression ratio of 11:1, the unit produced a lively 183 bhp at 7,800 rpm. Visual differences to the standard four-cylinder racing engine were the installation of a horizontal cooling fan (like that of the Formula 1 engine) and 44 mm Weber carburettors.

The Elva Mark VII chassis had to be altered to take the Porsche engine, of course, and this entailed moving the petrol and oil tanks, modifying the cockpit and, in particular, extending the rear part of the tubular frame. Since the Elva was fitted with 13 in diameter wheels (front rim width 6 in, rear 7 in), the five-speed Porsche racing gearbox also needed to be adapted. Thus modified, an Elva-Porsche accelerated from rest to 62 mph (100 km/h) in approximately five seconds and reached about 160 mph (260 km/h). Several American customers took delivery of Elva-Porsches—each with a price tag of over $10,000—in time for the 1964 racing season.

Also in 1964, a very light Elva Mk VII appeared in the European Hill-climb Championship, with a rear-mounted Porsche eight-cylinder racing engine (Type 771). Porsche's own hill-climber was now rather heavy compared with factory cars from Ferrari and BMW, and the factory was pessimistic about its chances, and as it turned out Edgar Barth, driver of the Elva-Porsche, won the opening event at Rossfeld beating Herbert Müller who was driving Barth's regular works Spyder (Type 718).

Despite this initial success it was felt that the Elva chassis was not strong enough to take the power, so Barth went back to his Spyder for the rest of the season while Müller drove the Elva-Porsche, and they captured the championship in that order. Down in seventh place in the table was Sepp Greger driving an Elva with a four-cylinder Porsche engine.

Above right *The Elva Mk VIII chassis was adapted to take Porsche engines.*

Right *With its four cylinders and 1.7-litre capacity, the Weber-carburated Porsche engine developed 183 bhp at 7,800 rpm.*

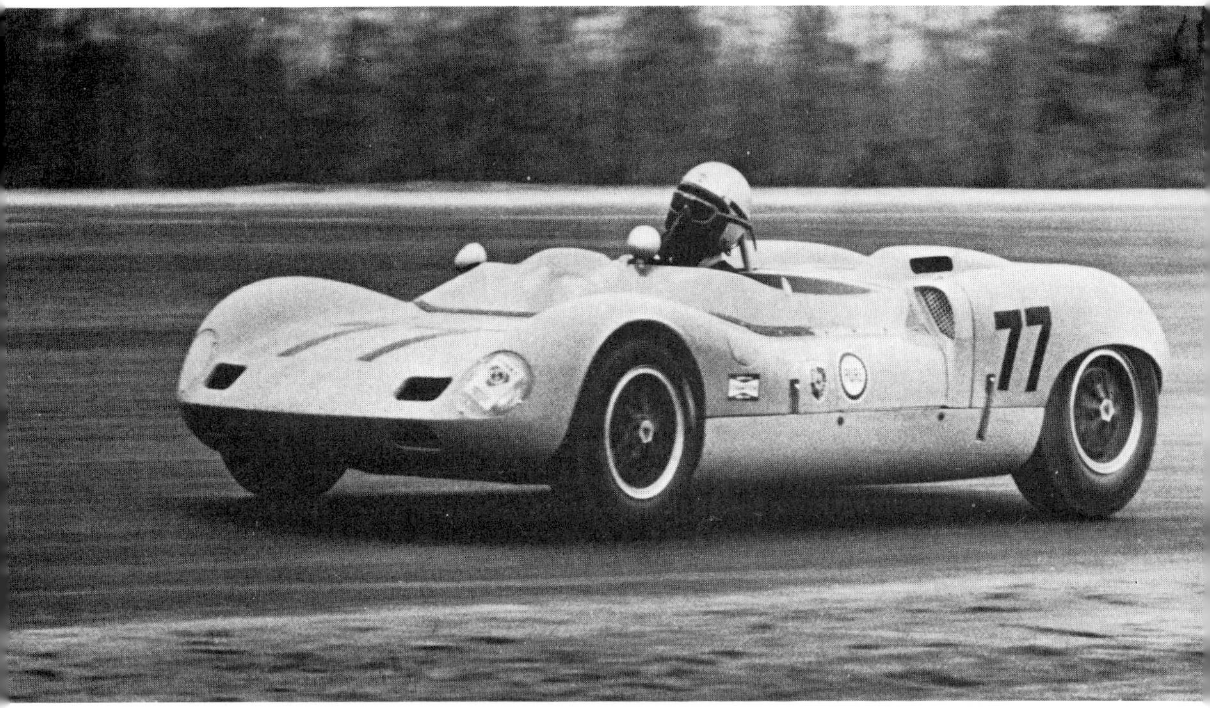
77

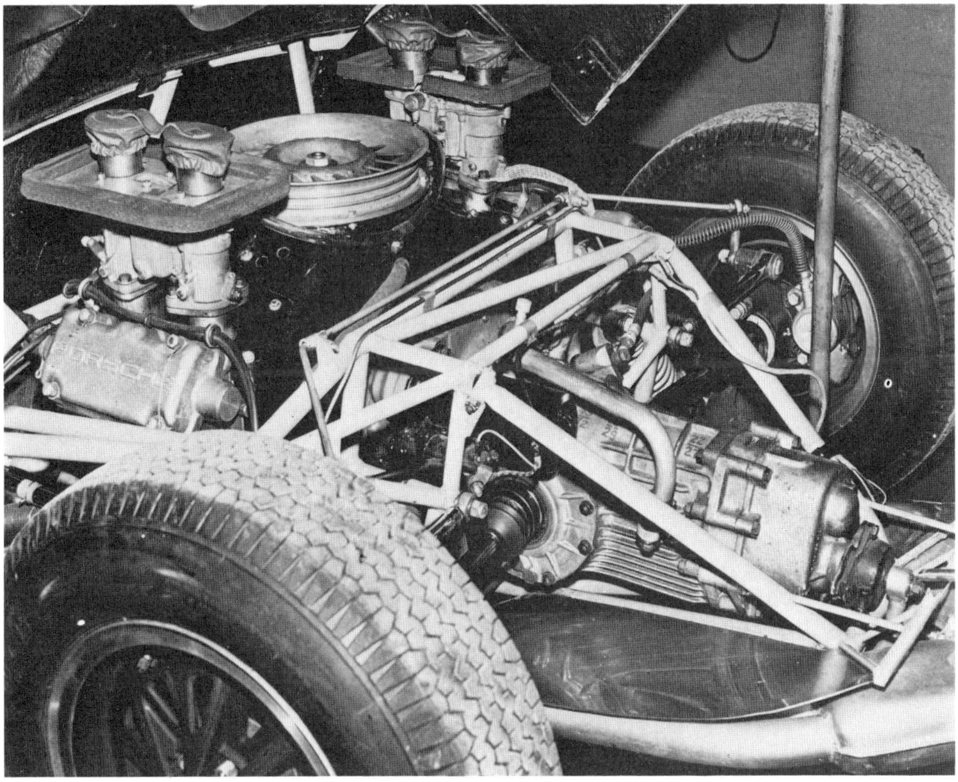

Emergency car: 924 base

To enable doctors to reach accident victims in the shortest possible time, Porsche introduced the 'Notarzt' purpose-built car, a 924, in April 1978. It was equipped with a two-way radio and car telephone, providing the team with full communications facilities.

Under the rear hatchback glass, which can be opened from the driver's seat, are a medical bag, a constantly charged battery, ECG unit, defibrillator and hand lamp. The equipment is easily accessible and can be used away from the car. The Notarzt 924 can be converted back to normal specification very easily, making it suitable for doctors who want to use it for private motoring as well.

The car shown at the Wiesbaden medical congress in 1978 had already been in use in the Würzburg area for the previous 21 months. Its specification was the same as for a production car, with a 0-100 km/h (62 mph) time of 9.8 seconds, a top speed of over 125 mph (200 km/h) and a turning circle of 33 ft (10 m). The main features are the distinguishing paintwork, two rotating lights on the roof, and horns under the front bumper.

The Porsche 924 'Notarzt' emergency car, with easily removed special fittings.

Envemo: 356 replica

Replicas of the Porsche 356C, in coupé and cabriolet forms, have been offered by the VAG and Porsche dealers Strähle in Schorndorf (near Stuttgart) since November 1981.

This really excellent replica has been produced in co-operation with a well-known Brazilian car body firm. In both forms the replica bodies are made of plastic, and the mechanical parts are based mainly on the VW Beetle, also 'Made in Brazil'.

The Envemo replica is faithful to the original 356 and attention to detail is quite amazing. In standard form the 'Super 90' is fitted with a 1.6-litre flat-four developing 50 bhp, although at extra cost there is a 1.8-litre version, developing 75 bhp, which will accelerate this Porsche replica to 62 mph (100 km/h) in fourteen seconds, and peak at 105 mph (168 km/h). Both the coupé and cabriolet comply with all applicable European and national safety standards, and carry price tags starting at DM 32,000.

The replica Porsche 'Super 90' has a 50 bhp VW engine and is faithful right down to the smallest detail.

Evex: 911 base

Egon Evertz, the enthusiastic amateur Porsche racing driver from Solingen, wanted to produce individual cars to represent the ultimate in driving experiences. With a team of specialists to carry out his ideas, he had Porsche customers in mind, but the first car he made was a two-plus-two VW-based sportscar called 'Evex 1'.

The plastic coupé was built on Volkswagen chassis and running gear, including a modified 2-litre engine which produced 120 bhp. It was 155½ in (3,950 mm) in length, 45 in (1,120 mm) high and weighed 1,765 lb (800 kg), the top speed being 125 mph (200 km/h).

Despite Evertz' optimism, 'Evex 1' never got beyond prototype form, but it was the basis for 'Evex 2' which appeared in 1978 and was developed from Porsche components. The design and construction of the plastic-bodied car, on the 911's floorpan, was certainly along the same lines.

Evertz started with a Porsche 911 SC powered by a 3-litre, six-cylinder engine, and DM 71,000 covered the major modifications to the nose, wings and rear-end. If the car was supplied, the body conversion alone was supposed to cost only DM 22,000, although there was no limit to what could be spent on beautifying the interior with special accessories. Despite the fact that the design was considered successful, Egon Evertz did not manage to join the ranks of recognized car improvers.

The first attempt by Egon Evertz appeared in the mid-1970s as the 'Evex 1'.

Evex 1 was based on Volkswagen chassis and running gear, although the 2-litre engine was tuned to 120 bhp. The 1,760 lb (800 kg) car had a top speed of 125 mph (200 km/h).

Evex 2, which appeared in 1978, was based on the 911 SC and its 3-litre flat-six engine.

Major modifications to the front and rear bodywork were priced at DM 22,000.

Fageol: twin-engine 356

Of great technical interest was a car which appeared in Europe early in 1954, produced by the American industrialist Lou Fageol: a Porsche 356 modified to be powered by two engines! According to Fageol, the twin-engine Porsche, with a pair of 1.5-litre power units, should have better handling and cornering qualities, despite the increased power. The second engine was mounted in the front compartment of the Porsche 356, involving modifications to the bodywork.

This was not Fageol's first twin-engine car, as it happened. In 1946 he had built a twin-engine car for the Indianapolis 500, powered by Meyer-Drake 1.5-litre supercharged engines. The car actually made the front row of the grid, but was eliminated by an accident after a few laps.

In 1952 Fageol produced a special version of a Porsche, but the twin-engined car of 1954 was elegant, having a roadster body which could be closed by a detachable hardtop.

The front engine drove the wheels through Rzeppa universal joints, and Porsche hubs were also used as well as an appropriately modified differential (turned 180°). A pair of Porsche synchromesh gearboxes transmitted the power, while front and rear suspensions remained as original, though reinforced with double-acting telescopic shock absorbers.

Both engines were controlled by one throttle pedal, one gear lever and one clutch pedal, the synchronization of engine speeds resulting only from the grip of the tyres on the road surface. The wheels, incidentally, were fitted with 6.50 × 15 tyres.

The total power of the two engines was 144 bhp. The car, which weighed 1,650 lb (750 kg), was supposed to have a top speed of 127.5 mph (205 km/h).

Flat Profile: 911 Turbo base

Since mid-1981 the Porsche factory has offered a modified front end, called *Flachbau* (flat profile), for the 911 Turbo. The front wings are modified, and the bumper and spoiler are replaced by the new nose section. This contains twin headlights with a cleaning system, indicator lights and an air inlet which serves both the oil cooler and the condenser for the air conditioning.

The wings are made of zinc-coated steel, following the original material, while the front end is made of glass-fibre reinforced plastic (GRP). Added to this are widened door-sill panels, making an elegant transition between the front and rear wings, which are also made of zinc-coated steel.

Side air inlets in the leading edges of the rear wings are offered as an option; these are not only visually different, but duct additional air to the engine, rear brakes and tyres.

Those who dislike the twin headlamps in the bumper could alternatively order fitted pop-up headlamps located in the flattened front wings. This type of mounting is certainly not easy, since the motors and linkage take space from the luggage compartment. Naturally these modifications have been registered and have received official approval. Depending on how much work is done, the modifications cost between DM 20,000 and 30,000.

Kaufmann: Turbo front-end modifications with headlights in the front spoiler (1981).

Porsche-Kremer: the Cologne tuner, who successfully adapted the 935 racing car from the time the factory ceased development, made a road car version which looked similar (1982).

The flat-profile version of the Porsche style, priced from DM 20,000 (1983).

Frua (Hispano-Aleman): 914-6 base

On behalf of Verne Ben Heidrich, the Spanish Porsche importer, Frua built a prototype based on the VW-Porsche 914-6 for the Geneva Salon in 1971.

Heidrich planned to manufacture a small series of this 'Hispano-Aleman' with a price tag of about DM 35,000. It had a large glass area, conventional doors, and air intakes in front of the rear wheels.

However, as soon as the show opened the project became the subject of a dispute between the designer and the customer regarding the rights to the design, Frua claiming that the reproduction rights should not be handed over by Heidrich to the Porsche company. A series of legal battles ensued for the next five years, eventually settled in Heidrich's favour.

Based on the Porsche 914-6 was Frua's design, which appeared on the coachbuilder's stand at Geneva in 1971.

Garretson: 914-6 base

Located in the American state of Colorado, Pikes Peak is only one of many mountains in the Rockies. It is rather special, though, as, at 14,110 ft (4,301 m), it is the scene of the greatest and most unusual hill-climb in the world. Although the event has been held on more than sixty occasions, only a few Europeans have taken part. It remained the domain of American drivers until Michèle Mouton won it, driving an Audi Quattro Sport, in 1985. Virtually all the 'greats' have made at least one attempt to crack the record on the hill, which is 12½ miles (20 km) long and rises almost 5,000 ft (1,500 m).

In 1975 the Garretson tuning company entered for the first time with a special car using a tuned 914-4 engine. Although initial trials went well, the car had an accident during qualifying, which put it out of the event. Nevertheless, Garretson made up his mind to return in 1976 with four cars, all powered by the 914-6 six-cylinder engine.

Using a simple, rugged single-seater chassis, Garretson fitted each car with a 2,368 cc Porsche engine rated at about 200 bhp, using an 11.5:1 compression ratio. Although Bobby Unser Junior was fastest in qualifying, the next three times were posted by Garretson-Porsches.

Rick Mears won the Pikes Peak hill-climb for Garretson on 4 July 1976, with a time only about 18 seconds outside the official record of 11 minutes 59.9 seconds. His brother, Roger, dropped out with a defective prop-shaft, while the third team member, Garry L. Kanawyer, finished eighth, after having gearbox problems.

Glöckler: 356 base/engines

Immediately after the war Porsche had no great desire to re-enter competitions, although virtually standard 356s were run at Le Mans in 1951, and by the beginning of the 1950s customers and dealers started to become active.

Walter Glöckler, the VW dealer from Frankfurt, had raced a Hanomag before the war, and he gave his general manager, Hermann Ramelow, instructions to build a fast sports car based on the Volkswagen. This home-built VW had a 1.1-litre engine developing a modest 48 bhp, but it weighed only 981 lb (445 kg) and enabled Glöckler to win the 1,100 cc category of the German sports car championship in 1950.

At the time there were increasing contacts between Glöckler and Porsche, and a close partnership was formed, resulting in a 'Glöckler-Porsche' which was made for the 1951 season. The Frankfurt firm of Weidenhausen built the body and the whole car weighed just 992 lb (450 kg), powered by a 1.5-litre Porsche engine which produced 85 bhp at over 6,000 rpm. The specification was good, but poor handling and brakes marred the concept, and the car was sold to an American.

In time for the 1952 season a third 'Glöckler' was produced, based on Porsche 356 parts, and it was driven by Glöckler's cousin, Helm, who scored a first-time win in the 1952 Eifel race. In contrast to the two mid-engined predecessors, the third had its engine over the rear axle, not ahead of it, enabling the wheelbase to be shortened by 2 in (5 cm).

The 1.5-litre Porsche engine produced 86 bhp, running on an alcohol mixture, and this car too was sold in America. Two more Glöckler-Porsches were produced in 1953, the fourth in the series ordered by Hans Stanek and largely based on the 1951 car, powered by the 1500S Porsche engine. The last was a roadster built for Richard Trenkel, powered by a modified 1.1-litre Porsche engine giving 67 bhp on an alcohol mixture.

Encouraged by the successes of Glöckler and Ramelow, the Porsche factory decided to enter the arena. By 1952 there was a whole crowd of competitors, so for the Le Mans race in 1953 Porsche built two Spyders, Type 550. On that occasion, they were still powered by a near standard 1.5-litre engine, but later cars used the Fuhrmann-designed Type 547 engine. These cars were inspired by the Glöckler-Porsches and launched the company's run of racing successes.

The first Glöckler-Porsche was built in 1950, powered by a 1.1-litre engine (Type 369) which developed 65 bhp. The lightweight mid-engined Spyder could achieve over 130 mph (210 km/h).

Above *Glöckler-Porsche Number 2 was built in 1951, with a 1.5-litre engine (Type 502) which achieved up to 95 bhp.*

Right *Sold in America in late 1951, the second car continued its successful racing career.*

Above *The rear view was the most familiar to rivals for some years.*

Left *The cockpit of Number 2 was spartan, but very practical.*

Below *The road version of the Glöckler-Porsche with a 1.5-litre engine ahead of the rear axle.*

Goertz: 914-6 base

Two special designs on the VW-Porsche 914-6 platform were seen at the Turin Show in 1970: the Tapiro from Giugiaro (see page 00) and the Eurostyle version designed by Count Albrecht Goertz.

Goertz, who had designed the BMW 507 Roadster and had also had a hand in the design of the Datsun 240Z, had high expectations for his design, since he knew that Porsche were not too happy about the 914's styling. The interior, and general engineering, were left original but the body style was completely changed, with a flattened nose and a roof line that continued to the back, where it fell away almost vertically—a Porsche 'station wagon'.

Although the Eurostyle 914 did not look too bad, and was well received in Zuffenhausen, it was not thought worthwhile at this point to alter the already well-established design of the 914-6. Goertz retained the car for a while, and later it went into a museum.

The Goertz Eurostyle 914-6 made its debut at the Turin Show in 1970.

Reminiscent of a station wagon, the Goertz 914-6 remained a one-off.

A completely new profile for the Goertz-designed 914-6 looked efficient.

Gordini: 550 engine

German-born Carl Delfosse emigrated to Argentina in 1952, and began to build up an efficient workshop for European cars. In the early days, Delfosse, who had won races in his homeland, bought a second-hand Gordini 1.5-litre racing car which had quite a history: Juan-Manuel Fangio had raced it in 1947.

After five months of construction and restoration a 'Gordini-Porsche' emerged from Delfosse's workshop, retaining the French chassis but clothing it with a new Spyder body. It was powered by a 1.5-litre Porsche engine, originally from a 356, but later this was changed for a more powerful 547 type which Delfosse bought from the well-known Porsche racing driver, Juhan.

Delfosse drove his successful hybrid to third place in the 1,500 cc class of the 1,000 km Buenos Aires race in 1955, then fitted a new body and made other modifications. The Gordini-Porsche weighed only 1,100 lb (500 kg) and, as Delfosse himself said, it was difficult to beat on twisty roads, although it was no longer competitive on high-speed circuits.

The unique Porsche-Gordini with spyder bodywork from the German-born Argentinian Carl Delfosse.

GRIPS: street-cleaning study

Many unusual projects have been undertaken by Porsche's research centre, but the commission from the German Federal Ministry for Research and Technology late in the 1970s was far from the usual run of things: it was to design a low-noise system for cleaning streets where parked cars were in the way, and the project was given the acronym GRIPS.

The problem was analysed by Porsche's best brains, and broken down into individual tasks. Their solution was to design a machine with an extension nozzle, which could clean underneath a parked car and retract again. The system involved rigid containers which could be rolled up by means of sprocket chains, with manual control, while a central suction fan performed the cleaning task. Sensors on the nozzles prevented them from becoming jammed underneath parked cars, making them retract automatically.

The study also put forward proposals for reducing noise on existing street-cleaning machines, and factors to be taken into account in their design.

Hess: 356 motor cycle

Peter Hess, an instrument mechanic from Kiel, undertook the task of building a motor cycle around the Super 90 Porsche 1.6-litre four-cylinder engine. History records that it took him three and a half years to build the machine, which cost DM 20,000 and a few grey hairs too!

The cooling fan had to be jettisoned and replaced by two oil coolers in tandem which increased the oil capacity from 3.5 to 6 litres.

To prevent the exhaust valves of the rear cylinders from overheating, they were given additional oil cooling. With original carburettors the speed of the engine was limited to 4,500 rpm, but there was still a good 60 bhp available, enough for a top speed of 112 mph (180 km/h). Without the engine speed limiter, it is estimated that the engine would have produced 100 bhp at 6,500 rpm, giving the machine a top speed of 150 mph (240 km/h)!

The engine's torque was impressive, drive to the rear wheel being via a BMW shaft. The forks were also from BMW, while Hess constructed the frame himself.

Hirondelle (Swallow): 356 engine/chassis

Dutchman Henke van Zalinge built the 'Hirondelle' (Swallow) spyder in 1957, and it had some very unusual technical features. The chassis was a tubular frame using VW and Porsche parts, the 1.6-litre Carrera engine was located at the front, and springing was performed by rubber components!

The car was raced in Holland for five years by such well-known drivers as Gijs van Lennep, Henk van Zalinge and Rob Slotemaker. Later in its life the Hirondelle was powered by a Porsche Super 90 engine.

Above The Dutchman, Henk van Zalinge, conceived the Hirondelle with a front-mounted Carrera engine.
Right The constructor van Zalinge raced the Hirondelle from 1957 onwards, as did Gijs van Lennep and Rob Slotemaker.
Below The front engine can be seen clearly in this view of the cockpit.

Hovercraft (Bonner): 356 engines

Three Porsche 1.5-litre engines from the 356 model, each rated at 52 bhp, were used to power a hovercraft built by the Australian, Colonel Colin Bonner, early in the 1960s.

His craft, named the 'Bon Air 6-1-X' was demonstrated to the Australian authorities and to the Army, and the Porsche engines performed very satisfactorily. Two of them blew air underneath the craft at about 1,500 cc per minute, to give it lift, and the third was used to power the propeller and the steering.

During the trial runs, with four people on board, the 'Bon Air' reached 56 mph (90 km/h) over water. It was made mainly of plastic and weighed 2,200 lb (1,000 kg), carrying a load of 1,200 lb (545 kg). It was 26 ft 3 in (8 m) long and 12 ft 6 in (3.8 m) wide.

In ideal operating conditions this one-off craft could carry up to eight people for a maximum of 90 minutes.

The 'Bon Air 6-1-X' built in 1963 was powered by three Porsche industrial engines, each rated at 52 bhp.

Hydrobus (hydrofoil): 356 engine

Originally designed as a speedboat, this hydrofoil built at the Lürsen dock, Bremen, in 1951 was the result of drawings by the engineer Schaper. He decided to increase its size and make it suitable for transporting passengers in coastal waters. The original power unit was a 1.3-litre Porsche engine, but later models also used 1.5-litre engines from Zuffenhausen.

The hydrofoil designed by engineer Schaffer used 1.3- and 1.5-litre Porsche engines. Designed as a speedboat, it was reworked to carry passengers.

Industrial engine: Type 616

The air-cooled VW engine was successfully offered with a continuous output of 22 bhp from 1952 (25 bhp from 1955), but it became clear that a Porsche engine would be required for higher-powered units. It helped that the Porsche engine, originally derived from VW parts, had the same overall dimensions.

By agreement with the VW factory, Porsche produced a 1.5-litre T engine in 1953 matching the VW unit insofar as design and special attachments were concerned; its power, however, was rated at 43 bhp at 3,600 rpm.

From the autumn of 1954 Porsche supplied their industrial engine, as well as their car engine, with a 1.6-litre cubic capacity (Type 616), and all further developments were also transferred to the industrial engine. The designs were basically the same, but in order to keep the thermal loadings within acceptable limits in continuous stationary operation, the engine was restricted to 43 bhp at 3,600 rpm, compared with 60 bhp at 4,500 rpm for the almost identical car engine. Apart from the speed limitation, the Type 616 also had its compression reduced from 7.5 to 6.5:1, and was equipped with only one carburettor. Without its speed limiter, the Type 616/3 produced 47 bhp at 4,000 rpm for fifteen minutes.

Fuel consumption ranged from 0.55 gal (2.5 litres) per hour at 2,500 rpm without load, to 3.41 gal (15.5 litres) per hour at 3,600 rpm under full load. The stationary engine differed from the car engine in the following main respects:

Battery ignition was replaced by a high-voltage magneto ignition, with speed limiter. Since the power unit was hand-cranked to start, with a handle, the battery, dynamo and starter motor were omitted.

The flywheel governor acting on the carburettor butterfly valve kept the engine speed constant.

There was only one carburettor and no fuel pump, since gravity from the tank to the carburettor was sufficient.

A simple safety cut-out developed by Porsche in co-operation with Bosch stopped the engine immediately if the fan belt broke.

As in passenger car applications, the Porsche industrial engine was mounted as a self-supporting unit on the crankcase flange. The mounting was similar to Volkswagen's so that the engines were practically interchangeable.

The 616 could also be adapted to run on gas instead of petrol. After appropriate tuning a power reduction of 6 per cent could be expected, but this reduced the operational costs by 32.5 per cent. In ten hours' full-load operation, at 3,000 rpm, some DM 30 could be saved.

The versatile Porsche industrial engine had many potential uses: portable fire extinguishers, power generator, automotive threshing machine, mobile cement silo, welding generator, mobile cable winch (up to 20 tons), rotary snow-plough, orchard sprayer, auxiliary drive, compressor or conveyor.

The 616/18 industrial engine followed early in the 1960s, with a continuous rating of 46 bhp against the 43 bhp of the 616/13. The peak power, for a maximum of fifteen minutes, was 55 bhp at 4,000 rpm without the speed limiter. The more powerful engine differed in that it had twin downdraught carburettors, a 7.5:1 compression ratio, twin exhaust pipes, was 4¾ in (12 cm) wider, and required higher quality fuel (86 RON).

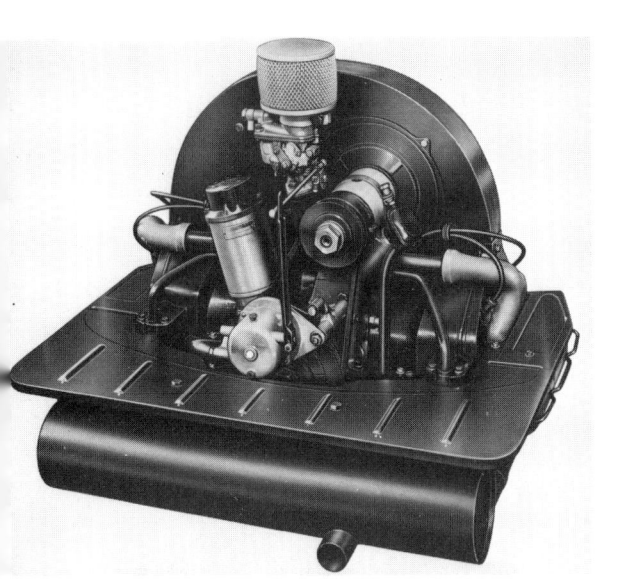

Above left *From the autumn of 1954, Porsche supplied the industrial engine Type 616 rated at 43 bhp.*

Above right *The 616/3 version with a flanged portable transmission.*

Below left *The mid-1960s' version was the 616/33-1, rated at 46 bhp.*

Below right *Industrial engine with magneto ignition and governor (standby generator).*

KMW-Sport: six cylinders + turbo

When the Porsche 917 Spyder first appeared early in the 1970s, private designers were tempted to enter the sports racing scene with variations on Porsche themes. One such many-faceted character was the designer Jo Karasek, whose attempts included both formula and sports car categories.

One of his designs was the 'KMW SP 2' sports car, a diminutive machine with a total length of just 140 in (3,500 mm) and a weight of less than 1,100 lb (500 kg). Its aluminium monocoque apparently had space for engines up to 5 litres, but on its debut the SP 2 had a 2-litre Porsche six-cylinder engine installed, developing 230 bhp. It was, of course, no match for the 5-litre, twelve-cylinder 917 which developed about 600 bhp.

Karasek's rolling chassis (without engine, gearbox, tyres or paintwork) cost a reasonable DM 26,800, and more powerful versions followed, notably with the 911R's four-cam six-cylinder, and later with the same engine plus turbocharger.

With the turbocharger the KMW went like a rocket in the early laps, often showing the way to the normally aspirated 917 Spyders, but it rarely finished a race. Later, up against the 917-10 Turbo Spyders, developing over 900 bhp, it stood no chance at all.

Jo Karasek's KMW SP 2 sports car, which appeared in 1973.

On its debut the SP 2 had a 2-litre six-cylinder engine developing 230 bhp.

The rolling chassis—a simple aluminium monocoque—cost only DM 26,800.

The ultimate version was fitted with a turbocharged 911R four-cam engine.

Kremer Racing: 935 base

None of Porsche's competitions customers could claim more success than brothers Erwin and Manfred Kremer, who now also run a flourishing Porsche dealership in Cologne. Erwin raced 911s very successfully in the 1960s and early 1970s, then became an entrant, notably preparing the cars for John Fitzpatrick and Bob Wollek on the six occasions when they won the Porsche Cup between 1972 and 1981.

The Porsche factory ceased development of the 935 Group 5 racing car in 1978, having produced the 935/78 model with water-cooled cylinder heads. Norbert Singer's team raced the car at Silverstone, where it won, again at Le Mans where it retired, and allowed it a swansong at the Norisring, then put it into the museum! It was then left to Porsche's racing customers, such as Kremer Racing and Reinhold Joest, to fly the flag against increasing opposition from Lancia-Martini.

The Kremer brothers' first attempt at modifying a production-line racing car was also the most successful, for their 935 K3 was the outright winner at Le Mans in 1979 in the hands of Klaus Ludwig, and Don and Bill Whittington. Ludwig won eleven of the twelve German Championship races of 1979 in the K3, usually with a 3.2 litre fully air-cooled engine.

The principal modification carried out by the Kremers was the design of lighter, subtly more aerodynamic bodywork made from kevlar material, notably with aerodynamic fences along the tops of the front and rear fenders. Space-frame chassis carried the front and rear suspensions, further improved from factory specification with extra bracing, and the third major modification was the use of an air-to-air inter-cooler, as envisaged by the factory early in 1976 until the FIA put a ban on it. The Kremers found a way of fitting it in a legal fashion, and had the advantages of lighter engine weight and greater reliability.

The Kremer Porsche 935 K3 continued to be successful during 1980, especially in the American IMSA series. For the 1981 season the K4 model was a further development using a full spaceframe chassis and a number of features of the 935/78, including the lowered floor and overall line, the inverted gearbox and 3.2 litre engines (still fully air-cooled) developing up to 800 bhp. Such a car, owned and driven by John Fitzpatrick, finished fourth at Le Mans in 1982 behind the trio of new Rothmans-Porsche 956 models.

At this point the Kremers built their most daring Porsche Special, the CK-5 Group C car, though it met with little success. Realising that the new 956 model was both expensive and not available to customers until 1983, the brothers quickly prepared a spaceframe chassis that owed more to the 917 than the 936, its wheelbase extended to 2,660 mm (104.7 in, and almost exactly the same as the new 956's wheelbase, as it happened). A high-backed, unlovely body clothed the CK-5, which was powered by the familiar, supposedly reliable 935 engine in the interests of low cost and good fuel consumption.

On its début at Le Mans in 1982 the CK-5 was among the quickest in a straight line, reaching 207 mph on the Mulsanne Straight, but it lacked ground effect and was already an also-ran against the factory 956s. Ted Field, Danny Ongais and Dale Whittington retired early at Le Mans with an engine problem, then at Spa three months later Rolf Stommelen and Stefan Bellof displayed a good turn of speed before retiring with starter motor trouble. Two successes were achieved in the German national sports car championship by the Kremer CK-5.

The best result in endurance racing was achieved by Richard Cleare, the Englishman who bought the first CK-5 and took it to sixth place overall in the Silverstone 1,000 km in 1983, with Tony Dron co-driving. Both CK-5 cars ran at Le Mans, Cleare's lasting eight laps before retiring with battery trouble, the 'works' Kremer driven by Derek Warwick/Frank Jelinski/Patrick Gaillard retiring before quarter-distance with a blown head gasket.

By now the Kremer-Porsche team was competing with 956 models almost exclusively, and Cleare's CK-5 appeared only rarely in British WEC events.

Below *The Kremer-Porsche team's 935 K3 fought through to victory at Le Mans in 1979.*

Bottom *Two Kremer-Porsche developments: John Fitzpatrick's 935 K4 followed at Brands Hatch (1982) by Kremer Racing's new C-K5. They finished third and sixth, respectively.*

Landing-craft engines: Types 170/171/174/174A

The water-cooled engines used by the pioneer divisions of the German Wehrmacht for landing craft were very susceptible to breakdowns when used in muddy water, so in 1943 the German High Command decided to try air-cooled engines for these craft. The only air-cooled, lightweight engine in quantity production which came anywhere near meeting the requirements was the Volkswagen flat-four, which was far from having enough power to do the job.

Porsche decided therefore to increase the unit's performance by using a Roots supercharger, this version of the engine being Type 170. A modified cylinder head design was also produced, with a higher compression ratio, to raise the power by more conventional means, and this was numbered Type 171.

This version had the disadvantage that few standard parts were retained, so new manufacturing facilities would be needed. Another variation, with larger inlet valves and a higher compression, was given the type numbers 174 and 174A. With this engine the fully laden landing craft had a top speed of 12 mph (20 km/h).

All four types, developed in 1943, had the normal bore/stroke ratio of 75 × 64 mm for a capacity of 1,131 cc, and had a maximum engine speed of 3,500 rpm.

The 170 had a compression ratio of 5.8:1 and with the Roots blower produced 34 bhp, but it was much heavier than the others at 463 lb (210 kg). Type 171 had a compression of 6.2:1 and achieved 33 bhp, with a weight of 256 lb (116 kg). Type 174, with a compression ratio of 5.8:1, remained virtually standard with 25 bhp, and weighed 253 lb (115 kg), while 174A had a higher compression of 7.5:1 and weighed 260 lb (118 kg), producing 30 bhp.

Below left *The supercharged Volkswagen engine, 170-1, produced 34 bhp at 3,500 rpm.*

Below right *The landing-craft engine, Type 171-2, had a 1.1-litre capacity and produced 33 bhp. It was built in 1943.*

London-Sydney: 911 base

Although Porsche was not represented by a works team on the London-Sydney Marathon in 1968, cars were prepared for three customers. All were competing in 911 models: the 1967 European Rally Champion Sobieslav Zasada with Marek Wachowski, Terry Hunter with John Davenport, and Edgar Herrmann with co-driver Hans Schuller.

The 911s were fitted with reinforced bumpers, strong roll-over hoops and steel mesh 'drapes' over the windscreens, to protect the crews (and cars!) from birds and wild animals. A particular worry on this great adventure was the danger of collisions in Australia with kangaroos: a 'roo in a hurry is capable of speeds up to 45 mph (70 km/h).

The Marathon started in London on 24 November 1968, and covered close on 10,000 miles (16,000 km) in all. The route took in Paris, Turin, Belgrade, Istanbul, Tehran, Kabul, Delhi and Bombay, where the leading 60 cars and crews were loaded on to a ship which took them to Australia. A further 3,000 miles (5,000 km) across the great continent, often guided only by a compass, took the survivors to the finish in Sydney.

All three Porsches suffered from fractured brake pipes in Turkey, Hunter being forced to retire with a broken piston. Only Zasada and Wachowski reached the end, after many problems, finishing in fourth place.

Three Porsche 911s were prepared for customers competing in the 1968 London-Sydney Marathon.

As protection against wild animals, ram fenders were fitted at the front and steel mesh guarded the windscreen.

Equipped for all eventualities on the 10,000-mile route, the production engine was fitted with extended exhausts for deep-water crossings.

Long-life car (FLA): study for a car of the future

A study carried out by the 'Club of Rome' on the state of the human race came to a grim conclusion: the raw material and energy reserves of the earth are hardly sufficient to support the grandiose scale of production and consumption of automobiles into the 21st century. The environmental protection lobby, meanwhile, was pointing out that the mountains of scrap represented by 12 million old cars every year were becoming a major problem.

Porsche's research and development centre at Weissach had already, early in the 1970s, begun a study into the possibilities of adapting car development to these changed environmental conditions. One result of these studies was the 'long-life car research project' (FLA).

This started out from the premise that cars with a doubled life-span would reduce by half both the raw material requirements and the amount of scrap. From the technical point of view it looked as though this problem could be solved by building cars which would last at least twenty years instead of ten (the statistical average life). So the study concentrated on two crucial points: reduction of wear and tear, and selection of materials.

Structurally, long life can be achieved by mass-production cars with low specific performance by employing hydrodynamic starting clutches, contactless ignition systems, and other measures. Appropriate selection of materials means corrosion-resistant bodywork made of aluminium, stainless steel or plastic, ozone-resistant rubber bushings, and special alloy brake discs or simply scratch-resistant discs.

Cars would take on a new character to become strictly functional, rugged, purpose-built machines built to standards never attained previously.

The second aim in materials selection, according to Porsche's study in 1973, is more intensive recycling: this would increase the proportion of re-usable materials and reduce that of disposable materials. Recycling would also result in savings in raw materials and smaller scrap-heaps. The FLA also hinted at effective financial economies since, according to Porsche's prognosis, a car lasting twice as long would cost only about 30 per cent more.

Long-life cars of this type would certainly involve changes for the motor industry, as well as for the dealers and garages. As with the shipbuilding and aircraft industries, they would have to change to a considerable extent from pure manufacturing to reconditioning, and parts modernization. The result would be a shift from the application of materials to the application of human labour.

The result of the research, in the form of a skeletal car, appeared at the Frankfurt Show in 1973 producing an immediate response from other car companies, and wide publicity in the world's press. The main argument against the car, it turned out, was the proposal that the buyer would keep his car for twenty years and pay a 30 per cent premium for it, forgoing technical advances that would be made in the meantime.

Events in the next ten years overtook the study, which was in any case meant to be a long-term prognosis. Two world energy crises, a downturn in the world's economy, and scarcer and dearer raw materials have all become important factors.

Many people have asked: 'What happened to the Long-life car?' The answer is that Porsche has incorporated many of its features in its production cars with the extensive use of non-corroding materials, plastics and alloys, and even the high-performance models should have no difficulty in surviving twenty years of everyday use. Surprisingly, though, it is the former opponents of the concept who are now supporting Porsche's FLA study, and who are adopting many of the recommended measures on their production lines. The demonstration car of 1973 has served its purpose, and given a new impetus to automobile development.

Specification—Long-life car

The aim of this project can be summarized as follows:

a) Optimization of the design with regard to low stressing of all components, together with appropriate materials selection.

b) Prolonging the life of all components with a view to providing an annual mileage of 15,000 km (9,300 miles) for twenty years with a minimum of maintenance, parts replacement, repairs and overhauls.

c) Simple and cheap maintenance, parts replacement, repairs and overhauls.

d) Higher scrap values due to improved recovery of raw materials, thus relieving environmental pressures.

Engine

Four-stroke Otto-cycle engine, cubic capacity circa 2.5 litres, power approximately 75 bhp at 3,600 rpm. Maximum engine speed 4,000 rpm. Specific power, approximately 30 bhp per litre.

Additional features for longevity would include hydraulic tappets; generously sized bearings; highly efficient air and oil filtration; fast build-up to operating temperature; and corrosion-resistant exhaust system.

Gearbox

Semi-automatic transmission with three forward gears, and a wear-resistant torque converter; possibly a fully automatic gearbox would be employed.

Additional long-life features would include an oil supply diverted from the engine oil system to the converter for thermal stability and ease of maintenance; long-life, generous arrangement of the gear wheels; optimum sealing of the drive shafts; and large oil reservoirs for extended replenishing intervals.

Chassis

The chassis life of current production cars is already so long that a life of twenty years, or at least 185,000 miles (300,000 km) is already attainable. Additional aids to longevity would include improved sealing of lubricated parts; generously proportioned, low-loaded springs; high-quality shock absorbers; special alloy brake discs; corrosion-resistant brake pipes; radial tyres with deep, long-life profiles; and foam-filling of cavities.

Bodywork

Corrosion-resistant body of aluminium or stainless steel, possibly also a combination of chrome-steel underpan or aluminium alloy for the construction. Plastic could be used if sufficiently capable of recycling, and wear-resistant.

Other long-life measures might include the deletion of paint-destructive fasteners on the bodywork; interior trim changed to ageing-resistant materials, including adhesive joints; improved scratch-resistance of glass. Attention should also be paid to details such as locks, hinges, window winders, seat rails and so on.

Electrics
Electronic, contactless ignition system, high-capacity maintenance-free battery with improved recycling capability. Other features would include separation of the cabling into multiple looms for easy fitting and removal; aluminium rather than copper pipes for improved recycling capability; and silver-coated plug contacts.

The FLA 'long-life research project' was presented at Frankfurt in 1973.

According to Porsche, a long-life car could last at least twenty years. Improved materials selection would make the best use of scarce raw materials. These new ideas provoked intense public discussion.

Marine engine: Type 729

With the many successes of Porsche power units on land and in the air, it seemed a natural progression for an appropriately modified four-cylinder unit to be used as a marine engine.

The Type 729 motor boat engine never achieved the successes of vehicle or stationary power units, however. For test and demonstration purposes Ferry Porsche had one of these engines fitted into his own motor boat, in the mid-'50s, but there was a great problem in getting enough cooling air to the unit.

The Type 729 was introduced with a continuous output of 52 bhp at 3,600 rpm. It was suitable for inboard mounting, with outboard transmission, twin screws and enclosed exhaust system and silencer.

Sea rescue launches were fitted with the Porsche marine engine in 1959, and various manufacturers used the power unit, but it never became a great commercial success.

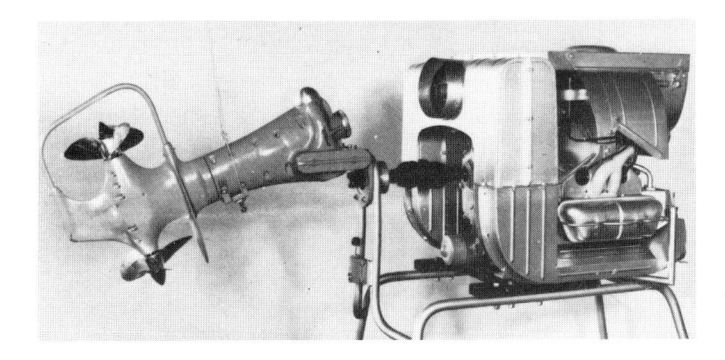

The Type 729 marine engine supplied a continuous 52 bhp at 3,500 rpm. The inboard-mounted engine had an external transmission and twin screws.

Mathé Universal: special single-seater

Otto Mathé, the one-armed Austrian racing driver from Innsbruck, and owner of the last remaining Berlin–Rome car—Type 64—built a single-seater from Porsche parts in 1952. For seven years he was among the fastest in European off-road events, on sand and ice.

With its even weight distribution allowed by the mid-engined configuration, Mathé's single-seater was more than a match for front-engined rivals.

A bulge on the bonnet hid the rev-counter, more visible there than inside the cockpit, and all the more important when a four-cam RS power unit was fitted. With this car Mathé won his last competition in 1959, an ice race against a works Spyder in the Porsche memorial meeting at Zell am See in Austria.

Mercedes-Benz land-speed record car: Type 80

The instigator of the T80 project—the Mercedes-Benz land-speed record car—in 1936 was none other than Hans Stuck, the Auto Union racing driver. At the peak of his racing career, Stuck still wanted more success and hatched the idea of achieving the world land-speed record for Germany—with himself at the wheel, of course.

At the time Malcolm Campbell held the record in 'Bluebird', which a 2,500 bhp Rolls-Royce aero-engine had powered to 301.2 mph (484.6 km/h). When Stuck began to get various people interested, Professor Porsche needed no persuasion; in fact, he was so enthusiastic that he designed the car free of charge.

The major problem was to find a suitable power unit which, according to Porsche, would need to develop at least 3,000 bhp. Daimler-Benz had been working on a new liquid-cooled V12 aero-engine since 1932, and by 1936 it was producing 1,300 bhp using a petrol injection system. Since the engines had been built as a secret project on behalf of the Reich air ministry, the government had disposition over them. Thanks to a good personal relationship between Stuck and the head of the supply office, Ernst Udet, approval was given for a pair of engines to power the land-speed record car.

After lengthy discussions, Daimler-Benz decided to back this project, with Porsche as their designer. In mid-October 1936 Stuck received written confirmation from Daimler-Benz for the financing of the project, while Stuck himself was to be responsible for the costs of the record attempt.

Early in November the Porsche design office received a general arrangement drawing of the new 'DB 601' aero-engine, which weighed 1,280 lb (580 kg) without a starter. When further trials proved that the Daimler-Benz engine could develop over 2,000 bhp using special fuel and high boost, Professor Porsche changed his original design and decided, in the interests of lower weight, to have a single engine, rather than two, powering the car.

On 22 March 1937 Porsche presented Daimler-Benz with the final design of the 'T80' project (this being Porsche's internal type number). With 2,200 bhp at 3,500 rpm the land-speed record challenger was to achieve its intended maximum speed of 342 mph (550 km/h) after a three-mile run-in. The shape of the T80 had been designed by Porsche to produce downforce on the front and rear axles by means of small aerofoils, and the drag coefficient was a sensational 0.18. To achieve this the underside of the car was completely enclosed.

In the meantime the performance of the Daimler-Benz aero-engine had been proven when, on 11 November 1937, a Messerschmitt Bf 109V 13 flew at 379.3 mph (610.4 km/h), a new world record. The power unit in the aircraft developed 1,650 bhp.

But the land-speed record was increasing all the time as George Eyston and John Cobb battled for the title. Eyston pushed his 'Thunderbolt' to 312 mph (502.1 km/h) at Bonneville, USA, using two aero-engines together producing 4,600 bhp. Cobb went faster but, on 16 September 1938, Eyston's speed of

357.5 mph (575.3 km/h) was the one that stood.

Porsche had to revise his plans yet again, for his original concept for the car allowed for a speed of only 342 mph (550 km/h). The new target of 373 mph (600 km/h) could be achieved if the T80 could accelerate for 4 miles, and have a further 1.5 miles in which to stop. Daimler-Benz found a more powerful aero-engine in the Type DB 603-V3 which could achieve 3,000 bhp, and this was fitted into the T80. The DB 603 had a cubic capacity of 44.5 litres (bore 162 mm/stroke 180 mm), the twelve cylinders arranged at a 60° V angle. Compression was just short of 7.5:1. Together with a flywheel weighing 161 lb (73 kg), the record engine weighed 1,780 lb (807 kg).

The T80 chassis comprised a space frame with side-members consisting of oval-shaped tubes, and the bare frame weighed 494 lb (224 kg). In order to accelerate the car as quickly as possible it was decided to use four-wheel drive, but the T80 was provided not with a gearbox, but with a so-called slipping clutch which locked up at around 93 mph (150 km/h), when the speed of the crankshaft matched the speed of the wheels.

In order to bring the car to a stop after just 1 km (1,000 yd) of braking distance, as calculated by Porsche, the T80 was fitted with six light-metal brake drums each 19.7 in (500 mm) in diameter. Each brake had four shoes, and three master cylinders had to operate 24 slave cylinders. The centre-lock wire wheels were 32 in in diameter and were fitted with 7 in wide Continental tyres. In high-speed tests the tyres were 'driven' at 420 mph (670 km/h) without any problems. Without fuel and bodywork the car now weighed 5,915 lb (2,683 kg), to which had to be added 494 lb (224 kg) for the tubular steel frame and 758 lb (344 kg) for the duralumin sheet metal body.

While Heinkel and Messerschmitt aircraft using Daimler-Benz engines had raised the air-speed record for piston engines to 470 mph (755 km/h), similar progress was being made at ground level . . . but without Daimler-Benz' presence. On 22 August 1939 John Cobb established a new land-speed record at 370 mph (595 km/h), and the T80's targets were getting harder all the time.

Other matters were pressing, and the idea of taking the T80 to America, where the records were being set, had to be abandoned. As a substitute the Reich Labour Service built a dead-straight track near Dessau, which was intended to be suitable for the world record attempt, but it was only just long enough, with nothing in reserve.

The T80, with its DB 603-V3 engine, was towed to the site on 12 October 1939. Minor problems were rectified immediately, but another month was needed to prepare it properly for its first record attempt. By now the war had begun and the project was abandoned. On 29 February 1940 the record engine was taken back to the factory and was used for aero-engine testing. The engineless T80 was taken to the factory workshop, and can be seen today in this form in the Daimler-Benz museum.

114

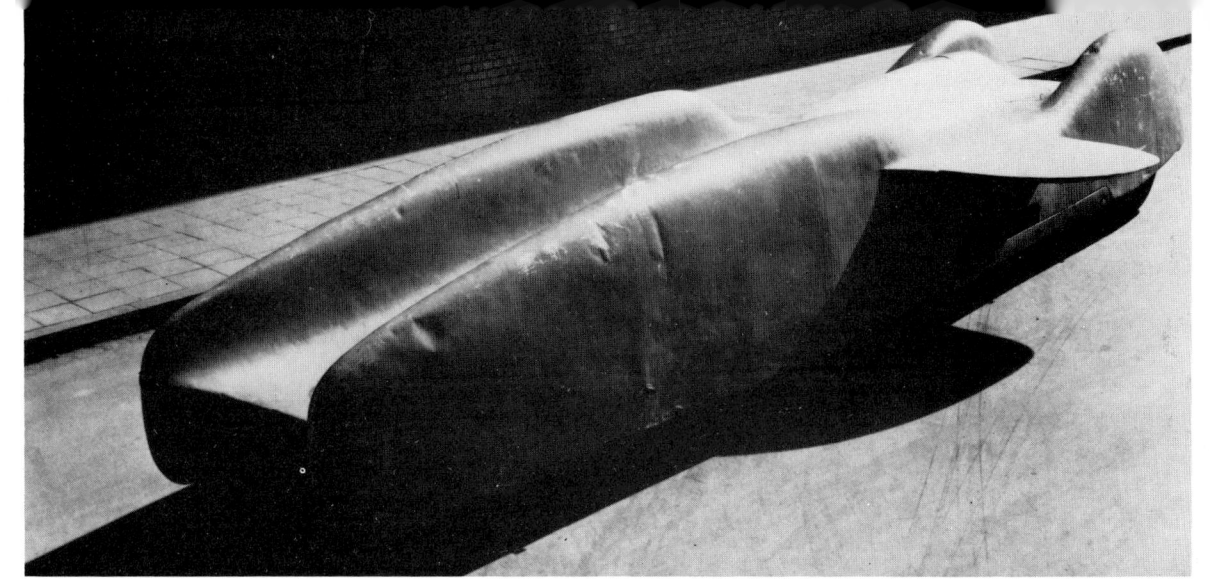

Left *The tubular space frame, which weighed 538 lb (244 kg), carried three axles, two of which were driven.*

This page *Daimler-Benz' 603-V3 aero-engine developed 3,000 bhp from a 44.5-litre capacity. The engine alone weighed 1,780 lb (807 kg), and the vehicle 6,950 lb (3,150 kg). The outbreak of war prevented a 370 mph plus (600 km/h) record attempt.*

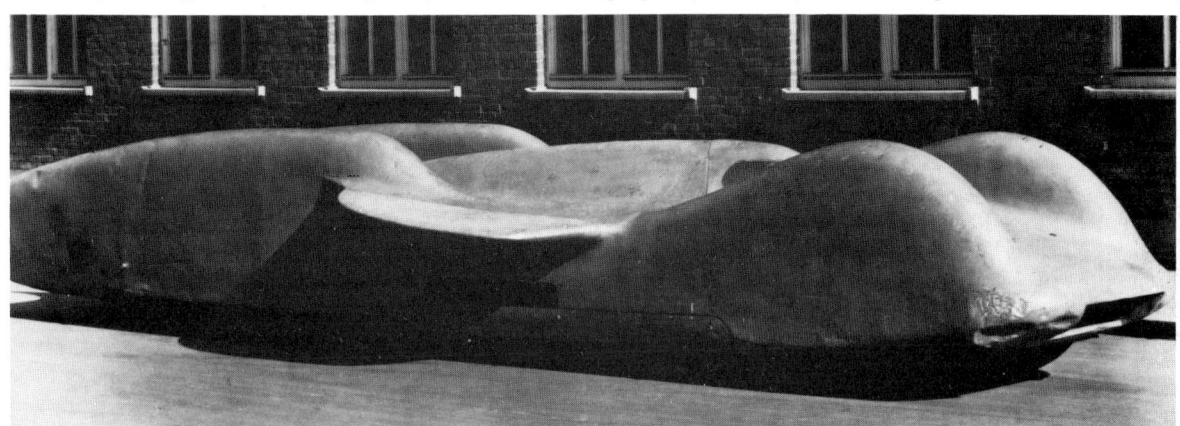

Mercedes-Benz racing engine: Type 94

In September 1936 the International Automobile Sports Commission announced the racing car regulations for the seasons 1938–40. For the first time designers would be given the free choice between normally aspirated and supercharged racing engines, the former with a maximum capacity of 4.5 litres, the latter with a limit of 3 litres.

Daimler-Benz produced two draft outlines for the new formula on 18 March 1937. During a round of discussions in the presence of Professor Porsche concerning the T80 land-speed record car, Porsche was asked for his opinion on the new racing formula.

By a stroke of luck Ferdinand Porsche just happened to have two design proposals in his briefcase. Like Daimler-Benz, he planned a V12 engine, although positioned at the rear following his concept of the Auto-Union, though Porsche now preferred to place the gearbox between the engine and the axle, providing better axle weight distribution.

While Daimler-Benz developed a supercharged 3-litre engine, Porsche was given the order to develop a 4.5-litre normally aspirated unit, though no one could tell which would be the better one for the 1938 racing season.

In November 1937 Porsche presented two draft outlines based on the eight-cylinder Wanderer engine: the W24. This radical design mounted three banks, each of eight cylinders, on a crankcase to form a W-shaped engine of 24 cylinders, the square dimensions of 62 mm × 62 mm producing a capacity of 4,490 cc. For this, Bosch were to supply three eight-piston injection pumps together with four six-cylinder magnetos. The engine allowed for only two valves per cylinder.

The second draft proposed a piston-valve control on an advanced two-stroke design. The results from the single-cylinder trials were disappointing, indicating a maximum of 305 bhp from the full-size design—much less than the 360 bhp that was anticipated.

In the meantime test-bed results for the supercharged 3-litre design had improved steadily, so Porsche's draft designs were rejected. The only item that Daimler-Benz wanted from Porsche was the injection system, but trials failed to show any advantages over normal carburation. Daimler-Benz even tried the Bosch injection system on the later 1.5-litre V8 engine, but still found no advantages, so further experiments were abandoned. Daimler-Benz engineers did return to injection in the 1950s, though, with considerably more success.

Midget: 911 engine

Midget car racing has been very popular in America for half a century, the single-seaters looking like racing cars of the 1940s, and powered by engines which must be located at the front, and not exceeding 2,276 cc in capacity. The centre of gravity therefore is unusually high by modern standards.

The mechanical design is very simple as direct drive is the norm, dispensing with clutch and gearbox. The Midgets have to be push-started, therefore, but reach a good 100 mph (160 km/h) on the quarter-mile ovals.

After years of domination by Ford V8 and Offenhauser engines, Larry Caruthers appeared at the beginning of the '70s with a VW engine in his Midget racer. The lighter engine and lack of water coolant reduced the weight and lowered the centre of gravity, which made it competitive. Well-known drivers, such as Gary Bettenhausen, raced VW-powered Midgets.

When the competition caught up again, the top runners switchd to the Porsche six-cylinder engine. Bettenhausen, for instance, built a 2.2-litre six-cylinder using a Hillborn methanol injection system and, although it was troublesome at first, the 200 bhp Midget-Porsche won a series of races in 1976.

The Midget racers in the States are frequently powered by Porsche's six-cylinder engine. Fuelled with methanol, the 2.2-litre units develop a good 200 bhp.

Mini-model engines: four/twelve cylinders

The Canadian engineer Herb Jordan, who worked for IBM before his retirement, undertook one of the most painstaking Porsche projects ever seen when he built scaled-down models of four- and twelve-cylinder racing engines. Jordan had raced a Formula 2 car with an engine he built himself in his younger days, but became hooked on Porsches when he bought a 356 Cabriolet and became increasingly involved with the Stuttgart marque.

Porsche supplied him with blueprints for the 904 four-camshaft engine, and Jordan built this on a reduction of 3:1. Since it was important that the engine should actually run, the engineer made minor changes as necessary. The Fuhrmann four-cam engine, originally with a bore/stroke of 92×74 mm (capacity 1,966 cc) was scaled down to 25.4×22.2 mm (capacity 45 cc). It took 2,000 hours to make, over a two-year period, but the Carrera engine started readily and, at 20,000 rpm, it produced eight bhp—177 bhp per litre!

With this project accomplished, Jordan set to work copying the 917's twelve-cylinder opposed engine on a reduced scale of 4:1, even using fuel injection and a turbocharger. Porsche's engineer Hans Mezger supplied the drawings.

The twelve-cylinder engine provided some headaches: for instance, to provide oil channels in the crankshaft, holes of only 0.00047 in (0.012 mm) were required. As Jordan said, it seemed desirable to keep friction low at 24,000 rpm, so he chrome-treated the crankshaft in order to avoid distortion during hardening. For the same reason the 7 in (17.8 cm) camshafts were machined in two sections. The camshaft housing aperture also had to be this length with a diameter of 0.0044 in (0.11113 mm), with a tolerance of ±0.000002 in (0.00005 mm)—this needed more than just a fine touch!

It took the Canadian four months just to make the injection pump. Here, too, micro-tolerances were needed for the six tiny pump pistons. Mini-injectors in the model 917 engine spray the mixture against the throttle slide, in the best Porsche racing engine tradition.

The 917 model started first time, thus beating the Carrera, which usually needed two attempts. To begin with the flat-twelve did not want to run up to its nominal speed of 24,000 rpm, and quite a lot of fine adjustment was needed to make it run to its design speed, but Jordan eventually succeeded. The unit has a capacity of 54 cc, or 4.5 cc per cylinder.

The Fuhrmann four-cam engine scaled down 3:1 by the Canadian engineer, Herb Jordan.

The second bit of magic—Jordan's twelve-cylinder opposed engine scaled down 4:1 (54 cc cubic capacity).

Moped engine: Type 655

Certainly the smallest engine design undertaken by Porsche was the 49.8 cc, four-stroke moped engine produced in the mid-'50s: the dwarf engine produced just 3 bhp at 7,000 rpm.

The suspended mounting, in which the cylinder head faced downwards, was a notable design feature, as was the design of the valve rockers. They were pressed out of steel plate and pivoted around a rounded impression on a corresponding track, which was secured to a pin.

The moped engine was provided with jet cooling, which consisted of two concentric pipes, the inner one being the exhaust pipe and the outer leading to the cylinder fins. The two pipes were so arranged that the exhaust gases, leaving the exhaust pipe at around the speed of sound, created an air flow in the outer pipe supplementing the normal air flow around the cylinder fins.

Another interesting feature is that in third (top) gear power was taken direct from the camshaft, which was connected to the crankshaft by a chain.

Murene: 914 base

The rush to produce special versions must have said something about the design of the VW-Porsche 914, and the French firm, Heuliez, presented its effort at the Paris Show in October 1970.

Jacques Cooper, an employee of the Heuliez company, which specializes in commercial vehicles and 'special projects', built the 'Murene' in just three months. The two-tone, mid-engined coupé remained much as original, although the bumpers were deleted, and the tail section could be let down in one piece. Access to the luggage compartment was provided by a separate lid.

Neumann: 356 aluminium roadster

Johnny von Neumann, Porsche's US West Coast distributor in the early years, bought from the factory one of the few remaining Gmünd-built aluminium coupés.

He had the roof sawn off, fitted a racing screen and raced the converted roadster successfully. It was still to be seen in competition as late as 1953.

Nordstadt: 911/928 engine

A real wolf in sheep's clothing appeared in Hanover at the end of 1972: a VW Beetle with Porsche Carrera running gear. Günther Artz, manager of the Nordstadt VW-Porsche agency, had commissioned the fastest VW ever built.

The chassis came from the 914-6, powered by a Carrera 2.7-litre engine and using many more Porsche parts. A VW 1303 body covered the 'special', which was equipped with racing wheels and high-performance tyres, and boasted a sporting interior. It took 2,000 hours to build the prototype, of which 1,000 hours were taken up just with welding.

The Beetle was a pure two-seater, since the six-cylinder engine took up a lot of space behind the front seats, but it had a top speed of 132 mph (213 km/h) and accelerated from rest to 100 km/h (62 mph) in just 7.3 seconds! The aerodynamics of the car were far from ideal, and the top speed was about 20 mph (30 km/h) slower than that of the original Porsche Carrera, although in acceleration to 100 km/h it was only a second slower than its illustrious cousin.

A real wolf in sheep's clothing: the VW Beetle with a Carrera engine, made by Nordstadt in 1972.
a *Steering: VW-Porsche 914* **b** *Tank: VW-Porsche 914* **c** *Dash panel: Porsche 911* **d** *Air vents: VW Variant 412* **e** *Front axle: VW-Porsche 914* **f** *Radial tyres: FR 70 VR 14* **g** *Rear axle: VW-Porsche 914* **h** *Additional oil cooler* **i** *Substructure: VW Porsche 914* **j** *Power unit: Porsche Carrera, 210 bhp, mounted ahead of rear wheels.*

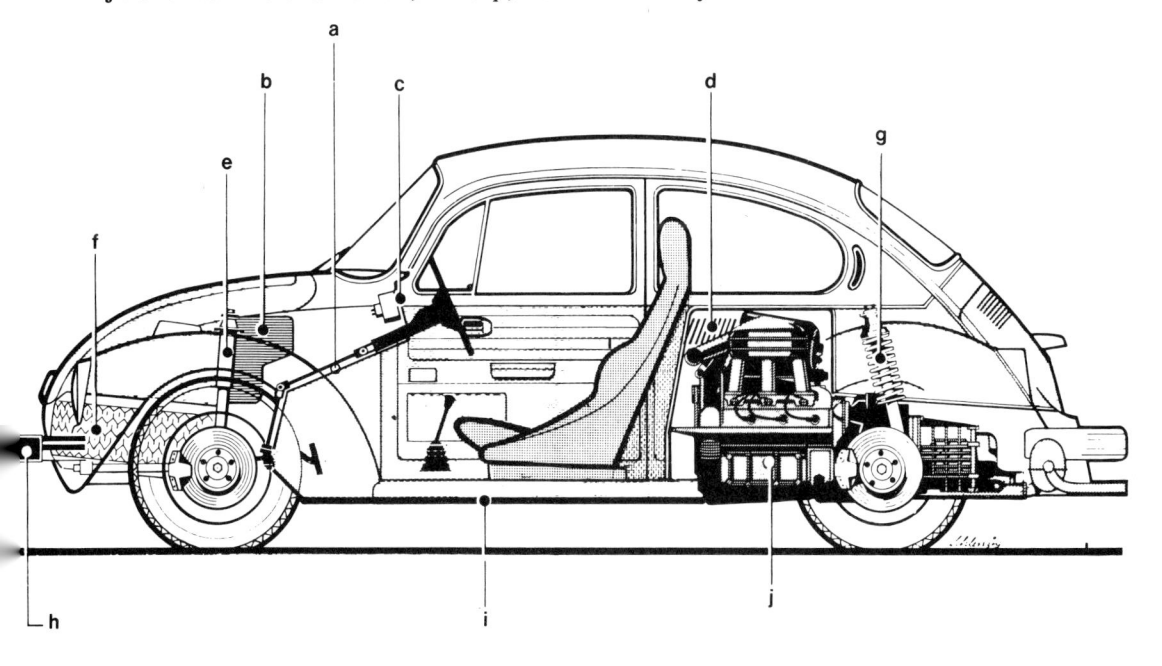

A standard-looking Golf body disguises Porsche 928 mechanics in Norstadt's special.

In 1979 Nordstadt produced a harmless-looking Volkswagen Golf—but the innocuous silhouette was deceptive. Nordstadt took the entire substructure of the Porsche 928—engine, chassis, gearbox and transaxle—and used the Golf body to disguise the power core. Actually the saloon had to be widened, lengthened and raised in height, too, to fit over the 928 chassis; and since it was not possible to incorporate the full range of instruments from the 928, expensive individual items had to be made.

Since the Golf-Porsche performed like a 928 it amazed every other road-user, as it could not be distinguished, at first glance, from a Wolfsburg product.

Ojjeh: 935 Street

Mansour Ojjeh, a leading businessman from Riyadh, in Saudi Arabia, contacted Porsche at the end of 1981. He was thinking of buying a Porsche, but it would have to be an out-of-the-ordinary modified Turbo.

As president of the TAG group (Techniques d'Avant Garde), Ojjeh was the financier of the V6 turbo racing engine for the McLaren Formula 1 team, and he had a clear idea of what his new Porsche should be.

A brand-new Porsche 911 Turbo went straight from the production line to the customer repair and preparation department at Porsche's Stuttgart-Zuffenhausen headquarters in February 1983. It was almost completely dismantled, and sheet-metal specialists then fitted a wide steel extension at the front of the chassis, while the plastics experts went to work at the back. Wherever possible off-the-shelf parts from the racing 935 model were used, notably for the double rear wing and the engine cowling.

It was still necessary to make such parts as the front spoiler, the bumpers and the door trim by hand—an expensive operation. The ride was kept comfortable by using Bilstein gas-filled shock absorbers, accompanied by stiffer anti-roll bars. The braking system was a combination of 911 Turbo standard parts and 934 racing car equipment. Centre-lock hubs were used on the extra wide wheels, which were fitted with Pirelli P7 tyres, 285/40 at the front and 345/35 at the rear. Even the spare wheel had to be made specially by BBS, since the centre-lock had to be taken into consideration, and there was only room for a 5½ in narrow rim, and motor cycle tyre, in the flat nose!

At the same time the engine department was busy modifying the power unit with a so-called 'power package', which raised the output by 70 bhp (to 370 bhp) without increasing the engine capacity or raising the normal 0.8 bar boost. The package included a larger air intercooler, a modified KKK turbocharger, higher lift camshafts, and a 934-type oil cooler in the nose.

In order to cope with the extra power, the four-speed gearbox was given additional lubrication and an extra oil cooler was fitted. While the exterior could be adapted with a number of existing parts, the interior decor required some special attention. For instance, a digital boost indicator, originally fitted in the racing Porsche 935, had to be adapted for the Ojjeh-Porsche. The interior was improved with a top-quality stereo system, leather upholstery, thick carpeting and a walnut dash panel.

Porsche's team worked a full three months on this Porsche special, which couldn't even be driven on German roads due to its non-standard specification. Mansour Ojjeh had fulfilled his dream of owning a roadworthy 935, and the car was handed over in July 1983. Though the price has never been revealed, a sum of DM 280,000 has not been denied. Exclusivity has its price!

Mansour Ojjeh, a Riyadh businessman, ordered a '935 Street' car from the factory. Based on a 911 Turbo, it was handed over in July 1983.

Centre-lock hubs were used, and the wheels were fitted with extra-wide Pirelli P7 tyres. The double rear spoiler, engine cowling and wings came from the 935 racing car.

The front luggage compartment could just take a specially made spare wheel.

The improved interior featured leather, walnut trim and additional instruments.

The specially modified 3.3-litre six-cylinder engine, with turbocharger, developed 370 bhp.

ORBIT: fire engine study/Type 2567

Each year in Germany fire damage claims over 1,000 lives, costs more than DM 3 billion, and causes inestimable lost production. In the continuing quest to reduce this toll, the German Federal Ministry of Research and Technology awarded a contract to Porsche requiring the research centre at Weissach to investigate how the use of fire engines could be improved, taking progress in vehicle technology into account.

The result of the study, which made the saving of life its priority, was published in August 1978. Porsche proposed a modular design for the fire-fighting system of the future, based on both existing production parts and on new components to be developed, using fire-fighting analyses for reference.

Three-dimensional view of the 'ORBIT' study, with one mode coupled.
a Forward seats b Driver c Breathing apparatus d Tank e Equipment storage, main vehicle f Equipment storage, supplementary module g Winch h Pump i Flow-control box j Generator k Coupling device l Flow-control box m Fire crew n Fire crew o Cooler.

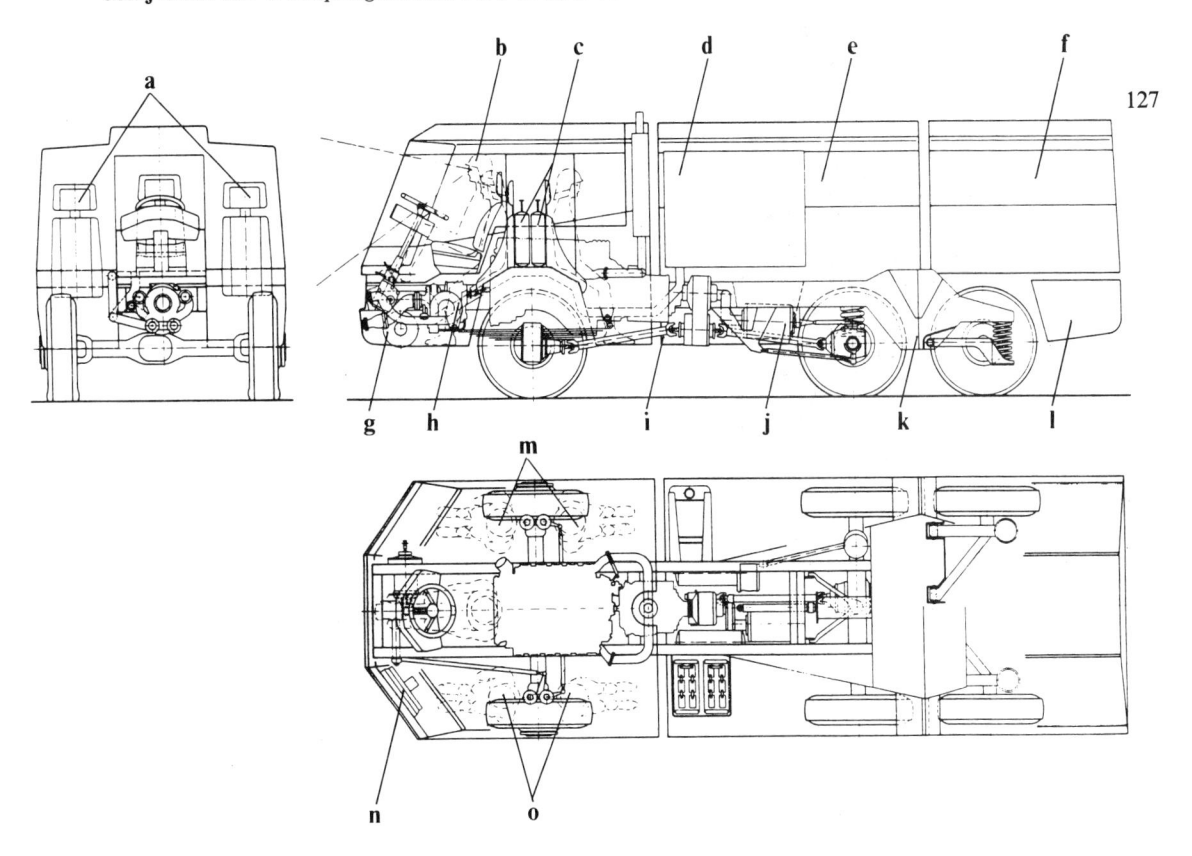

The components of the ORBIT fire-fighting system (the basis for the German acronym translates into Optimized Rescue and Fire-Fighting with Integrated Auxiliary Power) are a basic unit comprising chassis, driver's cab and body, extinguishing equipment and tools, together with a variety of modular extension units which are interchangeable but differently equipped. These modules can be exchanged or added on, on a building-block principle. The chassis acts mainly as a power source with drive motor, generator, winch and water pump.

Taking as an example a town with a population of 300,000 inhabitants, Porsche calculated that the investment required for equipping the fire service with conventional equipment would be in the order of DM 11.6 million. The new ORBIT system, however, would cost only DM 9.4 million, even though it provided greater cover. This would represent a cost saving of some 20 per cent, and, more importantly, a saving in terms of life and property.

ORBIT—study for a fire-fighting system (Type 2567), developed in 1978 as a modular concept.

Main unit plus supplementary trailer which, in this case, has an extension ladder.

The ORBIT system with main vehicle, extension ladder and supplementary module.

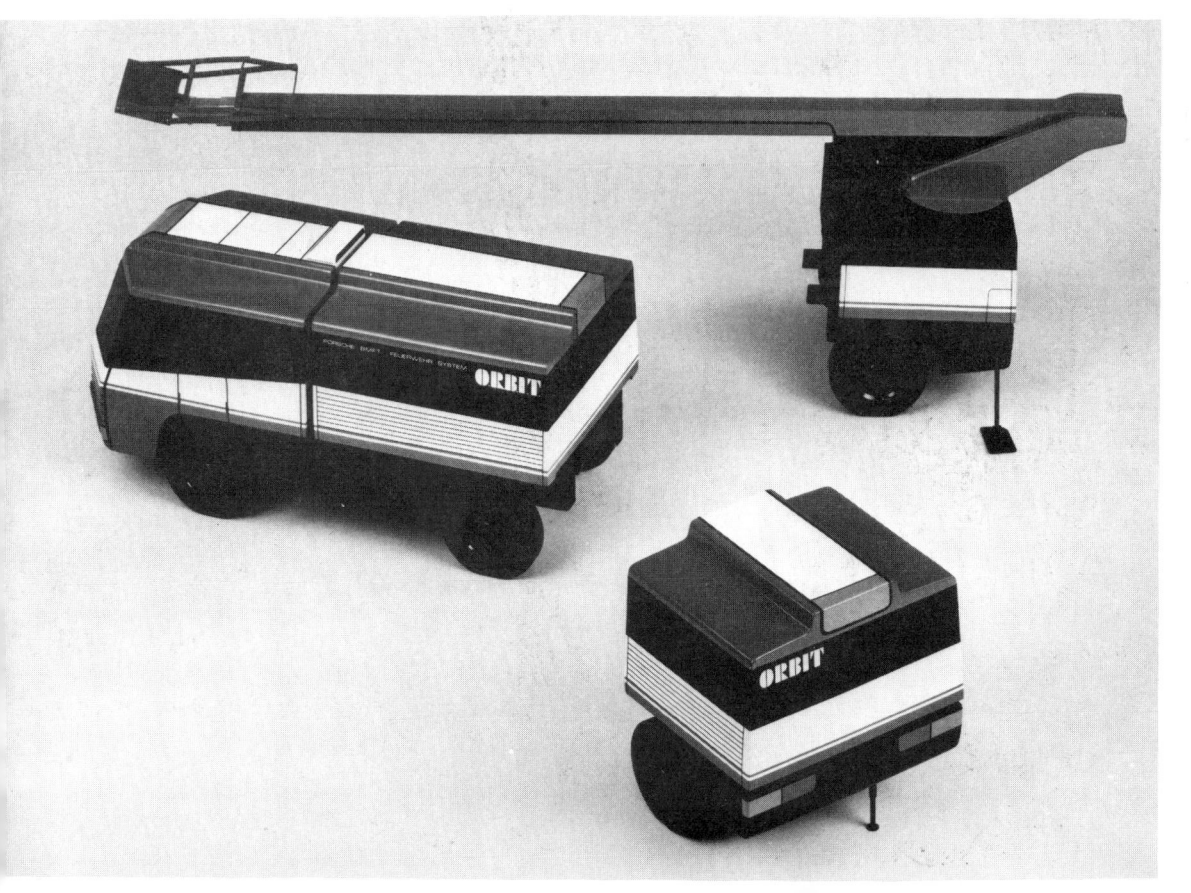

Ostrad traction engine: Type 175

As an alternative to the RSO tracked vehicle development from Steyr (some 16,700 were built, despite certain drawbacks), Porsche built the so-called Ostrad-Schlepper traction engine, its drawing board number Type 175.

Rather than tracks, as used on the Steyr, Porsche's giant had steel wheels with a diameter of 5 ft (1.5 m), and a width of 11.8 and 15.75 in (30 and 40 cm) respectively front and rear. The Ostrad traction engine was a four-wheel drive design, driven by a four-cylinder engine of 6-litre capacity, developing 80 bhp at 2,000 rpm. It was designed for use with a petrol, diesel or wood-gas engine (in the latter case, with supercharging!). To start the thing up half a VW engine (two cylinders, 12 bhp) was incorporated. At nominal engine revs the maximum speed was 9.3 mph (15 km/h)!

The Skoda factory in Pilsen began with a preliminary series of 100 vehicles early in 1943, of which about half were manufactured before the project was axed. According to Porsche, a total of about 200 engines (including test vehicles) was built, and production ceased because the disadvantages of steel wheels were too great.

Technical details of the Ostrad traction engine, Type 175, were as follows: the single or twin-plate clutch was supplied by Fichtel & Sachs; there were five forward gears and one reverse; the front-wheel drive was provided by bevel gears geared up 4:1, and the same applied to the rear-wheel drive; appropriate joints came from Rzeppa; the road wheels were an all-steel design with gripper strips, and the external turning radius at the front was 30 ft (7 m).

Porsche's Ostrad traction engine, Type 175. It had 59 in (1.5-m) diameter steel disc wheels, some 12 in (30 cm) wide. Power came from a four-cylinder multi-fuel engine with a 6-litre capacity, providing 80 bhp at 2,000 rpm.

Pick-up: 914-6 base

The idea of owning a pick-up truck with high performance came true when Dick Troutman from Costa Mesa, California, was offered a crashed VW-Porsche 914-6. The rear damage to this mid-engined two-seater was cut away up to the rear engine partition, and replaced with sheet metal to Troutman's design.

No changes were made to the rolling chassis. The small loading area ends in a hinged tailgate, and the work was carried out to a high standard. In fact Troutman, who owned a body repair shop, found that customers were queueing up for replicas. The six-cylinder Porsche 2-litre engine was not touched in this transformation, and the load area above it is surprisingly large.

A pick-up truck based on the VW-Porsche 914-6 was built in California by Dick Troutman. The chassis remained in its original form.

Pininfarina: 911 four-seater

Only rarely has Porsche approached a styling house for ideas, but this did happen when Pininfarina, the Italian bodywork specialist, was commissioned by the factory to make a four-seater 911. It appeared in the autumn of 1969, with the wheelbase of the 911 (chassis number 320020) lengthened by 7½ in (192 mm) to 97 in (2,460 mm). The chassis, transmission, clutch, and 2.2-litre six-cylinder engine developing 180 bhp, remained in standard form.

The four-seater, designated 'B 17' by Porsche, featured Boge Hydromat front suspension, and had shock absorbers from the same company at the rear. The anti-roll bars were ½ in (15 mm) in diameter, and the car ran on Michelin XAS tyres (185 VR 14) on 6-in wide Fuchs wheels.

At 2,500 lb (1,135 kg) the Pininfarina Porsche was really too heavy, and the weight distribution was badly impaired, with 39 per cent now on the front axle and 61 per cent on the rear, while the turning circle of 38 ft (11.7 m) was unacceptable, too.

In 1975 the one-off car was drastically modified. The bodywork was given the Carrera look with a front spoiler and wider wings, the latter to accommodate 7 and 8 in wheel rims carrying Dunlop 185/70 VR 14 and 215/60 VR 14 tyres. Koni shock absorbers were now fitted at the front and rear, and a 210 bhp Carrera 2.7-litre engine installed.

The four-seater Porsche 911, made by Pininfarina in 1969, was 7.5 in (19 cm) longer than the base model.

Above *The chassis, transmission and 2.2-litre engine (180 bhp) remained original.*

Left *Rear seat mouldings were unmodified, but the legroom was increased.*

In 1975 the Pininfarina 911 received many modifications and was given the Carrera look.

Wider Dunlop tyres were fitted to the 7 and 8 in rims under flared wheel arches. Koni shock absorbers replaced the Boge pneumatic struts.

The 1975 conversion included the installation of a 210 bhp, 2.7-litre Carrera engine to improve performance.

Police Porsche: 356/911/912/914/924

Police forces all over the world have ordered Porsches since 1956, to obtain the same advantages as those enjoyed by many thousands from the Zuffenhausen products. The cars had to be specially equipped, of course, for the surveillance of motorway traffic, accompanying important politicians, and for special duties.

The more complicated electrical systems and, more recently, electronics, to meet the demands of additional lights, headlights and sirens, have dictated the fitment of special generators capable of higher outputs. Radio communications systems with all ancillary equipment take up the entire boot space, and additional brackets on the roof are needed to mount the flashing blue lights.

In contrast to earlier versions, though, modern Police Porsches are quite inconspicuous, as much of the technology is hidden inside the cars.

Since 1956, in the 356 era, the popular models from the Zuffenhausen product range have been specially equipped for Police use.

Police forces all over the world respect the Porsche as a fast, reliable vehicle.

The 911, in coupé and Targa form, and the 912, 914 and 924, have been adapted for Police work.

Poll-Platje: 356 engine

Wim Poll, son of the VW dealer in Hilversum, built a spyder racing car in 1956 and won the Dutch racing championship several times. The mid-engined car had a VW swing axle, and Poll installed a Porsche Super 1300 engine in the tubular frame. The frame itself weighed only 99 lb (45 kg), and the whole car—called the Platje—weighed around 880 lb (400 kg). The fuel tank, in the nose, held 6 gal (28 litres).

The 1300 engine provided a good 67 bhp at 6,000 rpm with standard lubrication, and the car's maximum speed of about 125 mph (200 km/h) was reached in fourth gear. Later Porsche offered Poll a 1600 engine, which extended the car's run of successes.

This mid-engined spyder, powered by Porsche, was produced in Holland in 1956. Its constructor, Wim Poll, was several times the Dutch champion in the 'Poll Platje'.

Pooper: Formula 2

The South African driver, Peter Lovely, became well known in America in the mid-'50s, driving his 'Pooper'. Lovely fitted a streamlined body on to a Cooper F2 chassis and installed a Type 356 1500S engine. On several occasions the combination became a real embarrassment to the original Porsche Spyder (Type 550) teams.

The 'Pooper' was created by South African driver Peter Lovely, who installed a 1.5-litre Porsche engine in a Cooper F2 chassis and gave it special bodywork.

Porsche: Type 52 sports car

As a spin-off from the successful Auto Union Grand Prix car, Type 22, the Porsche design studio developed a road car for the Auto Union group: a sports coupé, powered by the successful sixteen-cylinder engine!

Two prototypes of the Type 52 were built in 1935, but nothing more than drawings and technical specifications exists today. The car had three seats abreast, with the driver located in the centre. The engine, ahead of the rear axle, was a detuned version of the sixteen-cylinder racing engine with 5-litre capacity (although another source mentions 4.4 litres), and was supercharged. The engine was quoted at 200 bhp, while maximum torque would be achieved at just 2,250 rpm.

The reinforced platform frame carried the running gear, all four wheels being independently sprung and damped by hydraulic shock absorbers. The five-speed gearbox was located behind the rear axle. The crankshaft ran in ten bearings, and a single camshaft operated all 32 valves.

Eight Solex carburettors, each of 1.9 in (48 mm) diameter, provided the correct mixture for the engine. At the back a triple-plate clutch was installed, transmitting power to the wheels. The car weighed 2,870 lb (1,300 kg) and had a top speed of over 118 mph (190 km/h)—sensational by the standards of 1935.

With a wheelbase of 117.6 in (3 m) and wheel rims of 20 in diameter, the Type 52 made an impressive sight with its teardrop aerodynamic body. From the front it looked much like the racing car, and used the same radiator. The side view revealed a high waistline and fashionably small windows, and the roof swept back in an elegant line straight to the rear. After examining the two prototypes, however, Auto Union decided not to proceed with the project.

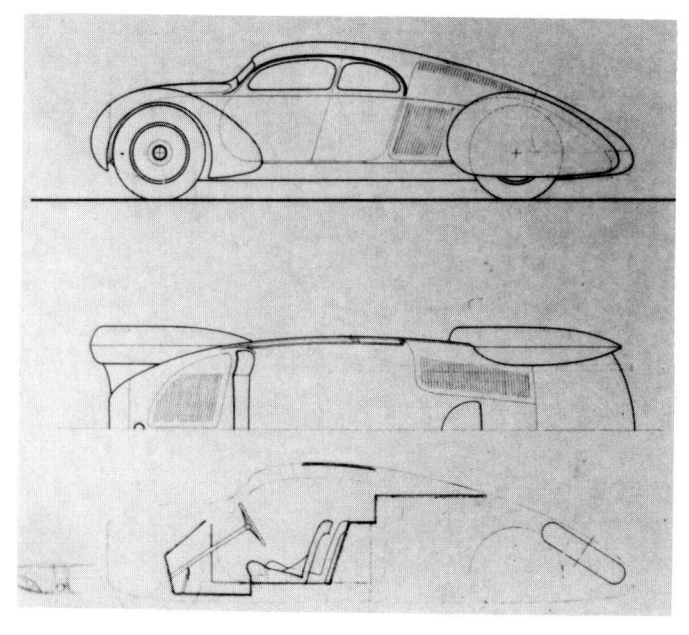

A super sports coupé was designed by Porsche in 1935, based on the sixteen-cylinder Auto Union Grand Prix car.

TECHNISCHE DATEN — Sportwagen Typ 52

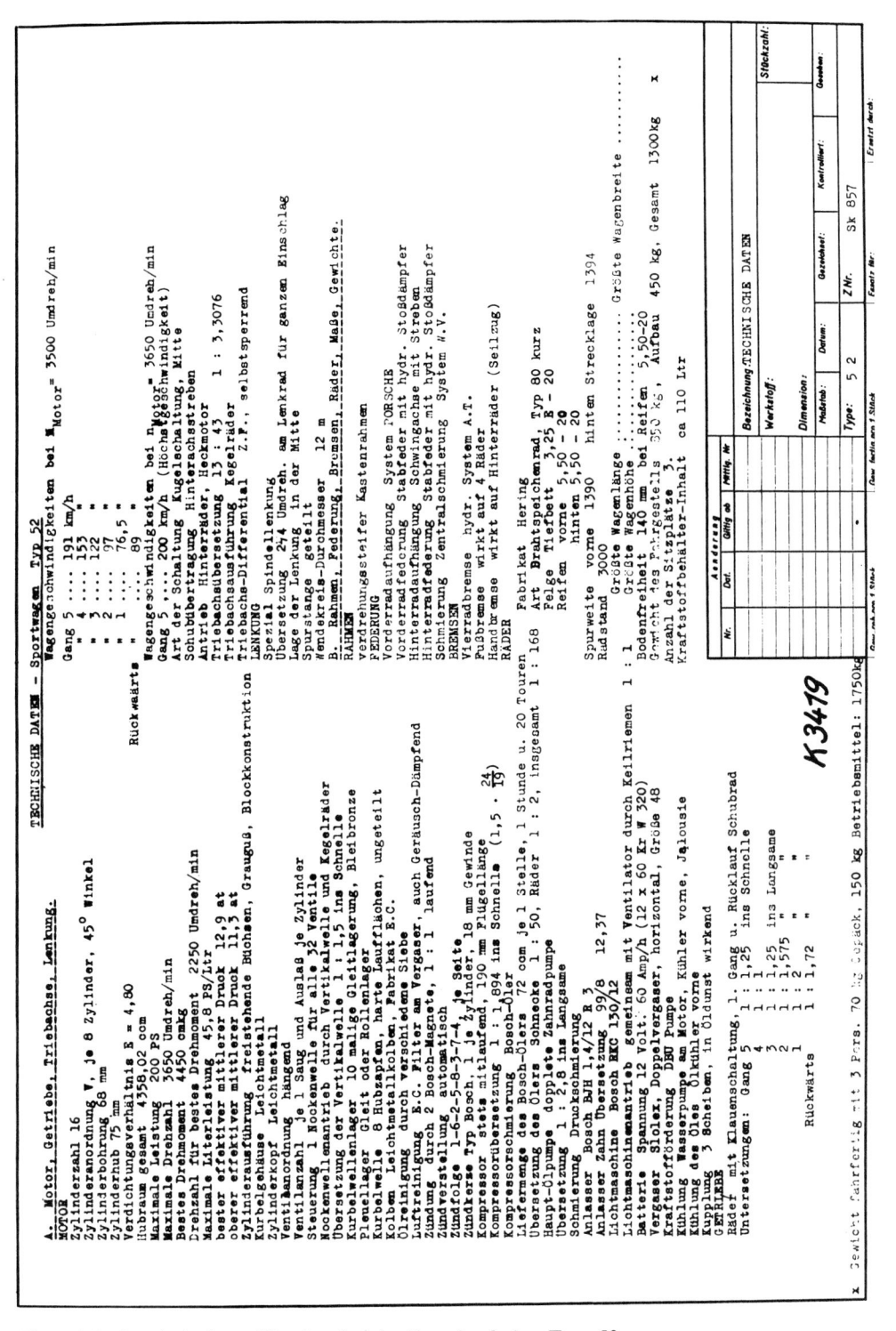

A. Motor, Getriebe, Triebachse, Lenkung.

MOTOR
Zylinderzahl 16
Zylinderanordnung V, je 8 Zylinder, 45° Winkel
Zylinderbohrung 68 mm
Zylinderhub 75 mm
Verdichtungsverhältnis E = 4,80
Hubraum gesamt 4358,02 ccm
Maximale Leistung 200 PS
Maximale Drehzahl 3650 Umdreh/min
Bestes Drehmoment 4450 cmkg
Drehzahl für bestes Drehmoment 2250 Umdreh/min
Maximale Literleistung 45,8 PS/Ltr
bester effektiver mittlerer Druck 12,9 at
oberer effektiver mittlerer Druck 11,3 at
Zylinderausführung freistehende Büchsen, Grauguß, Blockkonstruktion
Kurbelgehäuse Leichtmetall
Zylinderkopf Leichtmetall
Ventilanordnung hängend
Ventilanzahl je 1 Saug und Auslaß je Zylinder
Steuerung 1 Nockenwelle für alle 32 Ventile
Nockenwellenantrieb durch Vertikalwelle und Kegelräder
Übersetzung der Vertikalwelle 1:1,5 ins Schnelle
Kurbelwellenlager 10 malige Gleitlagerung, Bleibronze
Pleuellager Gleit oder Rollenlager
Kurbelwelle 8 Hubzapfen, harte Laufflächen, ungeteilt
Kolben Leichtmetallkolben, Fabrikat E.C.
Olreinigung durch verschiedene Siebe
Luftreinigung E.C. Filter am Vergaser, auch Geräusch-Dämpfend
Zündung durch 2 Bosch-Magnete, 1:1 laufend
Zündverstellung automatisch
Zündfolge 1-6-2-5-8-3-7-4,4° Seite
Zündkerze Typ Bosch, 1 je Zylinder, 18 mm Gewinde
Kompressor stets mitlaufend, 190 mm Flügellänge
Kompressorübersetzung 1:1,894 ins Schnelle (1,5 · 24/19)
Kompressorschmierung Bosch-Öler
Liefermenge des Bosch-Ölers 72 ccm je 1 Stelle, 1 Stunde u. 20 Touren
Übersetzung des Ölers Schnecke 1 : 50, Räder 1 : 2, insgesamt 1 : 168
Haupt-Ölpumpe doppelte Zahnradpumpe
Übersetzung 1 : 2,8 ins Langsame
Schmierung Druckschmierung
Anlasser Bosch BJH 1,4/12 R 3
Anlasser Zahn Übersetzung 99/8   12,37
Lichtmaschinenantrieb gemeinsam mit Ventilator durch Keilriemen
Batterie Spannung 12 Volt; 60 Amp/h (12 x 60 Kr W 320)
Vergaser Soles, Doppelvergaser, horizontal, Größe 48
Kraftstofförderung DBU Pumpe
Kühlung Wasserpumpe am Motor, Kühler vorne, Jalousie
Kühlung des Öls Ölkühler vorne
Kupplung 3 Scheiben, in Öldunst wirkend

GETRIEBE
Räder mit Klauenschaltung, 1. Gang u. Rücklauf Schubrad
Untersetzungen: Gang 5   1 : 1   ins Schnelle
                     4   1 : 1,25   "
                     3   1 : 1,575   "
                     2   1 : 1,28   ins Langsame
                     1   1 : 1,72   "
             Rückwärts

Wagengeschwindigkeiten bei M_{Motor} = 3500 Umdreh/min
Gang 5 ..... 191 km/h
  " 4 ..... 153 "
  " 3 ..... 122 "
  " 2 ..... 97 "
  " 1 ..... 76,5 "
Rückwärts   89 "

Wagengeschwindigkeiten bei n_{Motor} = 3650 Umdreh/min
Gang 5 ..... 200 km/h (Höchstgeschwindigkeit)
Art der Schaltung Kugelschaltung, Mitte
Antrieb Hinterräder, Heckmotor
Schubübertragung Hinterachsstreben
Triebachsübersetzung 1 : 3,3076
Triebachsausführung Kegelräder
Triebachs-Differential Z.F., selbstsperrend

LENKUNG
Spezial Spindellenkung
Übersetzung 2¼ Umdreh. am Lenkrad für ganzen Einschlag
Lage der Lenkung in der Mitte
Spurstange geteilt
Wendekreis-Durchmesser 12 m

B. Rahmen, Federung, Bremsen, Räder, Maße, Gewichte.

RAHMEN
verdrehungssteifer Kastenrahmen

FEDERUNG
Vorderradaufhängung System PORSCHE
Vorderradfederung Stabfeder mit hydr. Stoßdämpfer
Hinterradaufhängung Schwingachse mit Streben
Hinterradfederung Stabfeder mit hydr. Stoßdämpfer
Schmierung Zentralschmierung System W.V.

BREMSEN
Vierradbremse hydr. System A.T.
Fußbremse wirkt auf 4 Räder
Handbremse wirkt auf Hinterräder (Seilzug)

RÄDER
Fabrikat Hering
Art Drahtspeichenrad, Typ 80 kurz
Felge Tiefbett 3,25 B - 20
Reifen vorne 5,50 - 20
       hinten 5,50 - 20

Spurweite vorne 1390   hinten Streckslage 1394
Radstand 3000   Größte Wagenlänge   Größte Wagenbreite ....
                 Größte Wagenhöhe
Bodenfreiheit 140 mm bei Reifen 5,50-20
Gewicht des Fahrgestells 850 kg, Aufbau 450 kg, Gesamt 1300 kg
Anzahl der Sitzplätze 3.
Kraftstoffbehälter-Inhalt ca 110 Ltr

x Gewicht fahrfertig mit 3 Pers. 70 kg Gepäck, 150 kg Betriebsmittel: 1750 kg

K 3419

Nr.	Änderung	Def.	Gültig ab	Mittlg. Nr	Bezeichnung TECHNISCHE DATEN	Stückzahl:
					Werkstoff:	
					Dimension:	Gezeichnet:
					Modellgr. 5 Datum: 2	Kontrolliert:
					Type:	ZN. Sk 857

The original technical specification for the Porsche design Type 52.

Porsche: Type 514

When Porsche received an invitation from Charles Faroux to take part in the Le Mans 24-hour race in 1951, the company was keen to accept, though it lacked a suitable car to take on the specialized machines that would win the race outright. The decision was taken to run the aluminium-bodied 356 coupés that had been made in the Gmünd factory, to demonstrate the worth of the production cars.

Two coupés, later called the Porsche 356 SL (Super Lightweight) were to represent the Porsche marque on its return to France. The SL had slotted aluminium plates instead of rear windows, a petrol tank increased to 17 gal (78 litres) with an external filling nozzle, and covered wheels. The wheels had larger holes to help with brake cooling, and under the bonnet and tail aluminium plates smoothed the airflow. For Le Mans the 1,086 cc engines were mildly tuned to 46 bhp, allowing the 1,400 lb (635 kg) coupés a top speed of 100 mph (160 km/h).

Bad luck struck the works team before the race, both cars being badly damaged in road accidents. A reserve car was prepared as quickly as possible,

Le Mans 24-hour race in 1951: Porsche's 356 aluminium coupé (Type 514) won its class.

The 356 SL was powered by a 1.1-litre engine developing 46 bhp. It weighed 1,400 lb (635 kg) and had a top speed of 100 mph (160 km/h).

and the two damaged cars were turned into one good one. Even then one of them was crashed during practice, so only one Porsche took the start in the hands of Auguste Veuillet and Edmond Mouche. Without any more mishaps they finished twentieth, out of 29 classified, and won their class.

Porsche: Type 530 four-seater

Very soon after the 356 went into series production at Zuffenhausen, some thought was given to making a genuine four-seater version of the two-plus-two coupé. With the type designation 530, two prototypes were built in 1951/52 on a 94½ in (2,400 mm) wheelbase; one a coupé and one a cabriolet. Each was powered by a 1.5-litre four-cylinder engine.

On the coupé version the bumpers were still below the bodywork line. The roof was a 'notchback' design, in which the rear side windows followed the slope and could be opened to the rear. The back-seat passengers, who sat between the rear wheel arches, had rather better accommodation than in the normal coupé, and the front doors were made larger to facilitate access to the back seats.

The cabriolet had bumpers that were wrapped around the sides. The prototype had a folding roof, rear winding windows, and folding rear seats to increase the luggage space (the coupé also had this feature). From the technical point of view both prototypes corresponded to 1952 specifications. Ready to drive, they weighed 2,000 lb (910 kg). After extensive trials the design was rejected for series production, for marketing reasons.

Opposite page *A couple of four-seater prototypes (Type 530) based on the 356 appeared in 1951, but the project was not put into production for purely marketing reasons.*

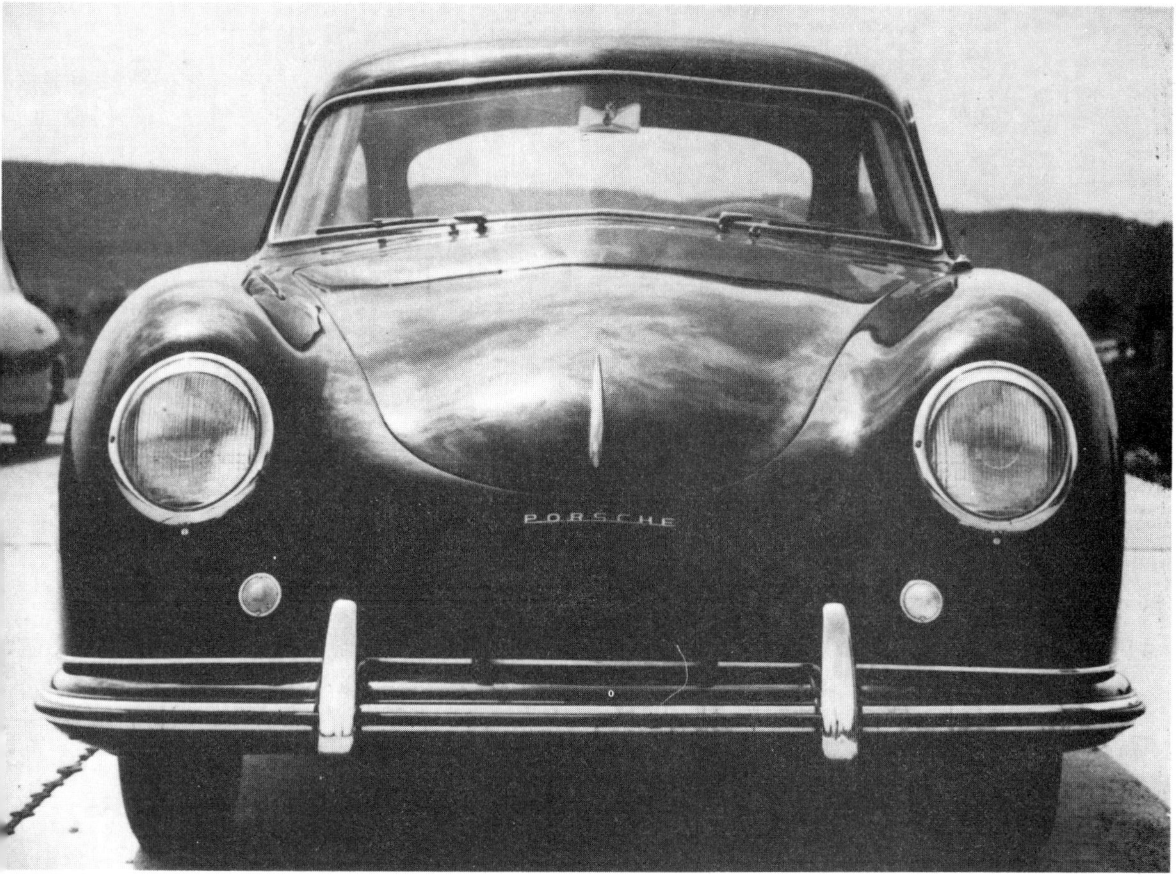
PORSCHE

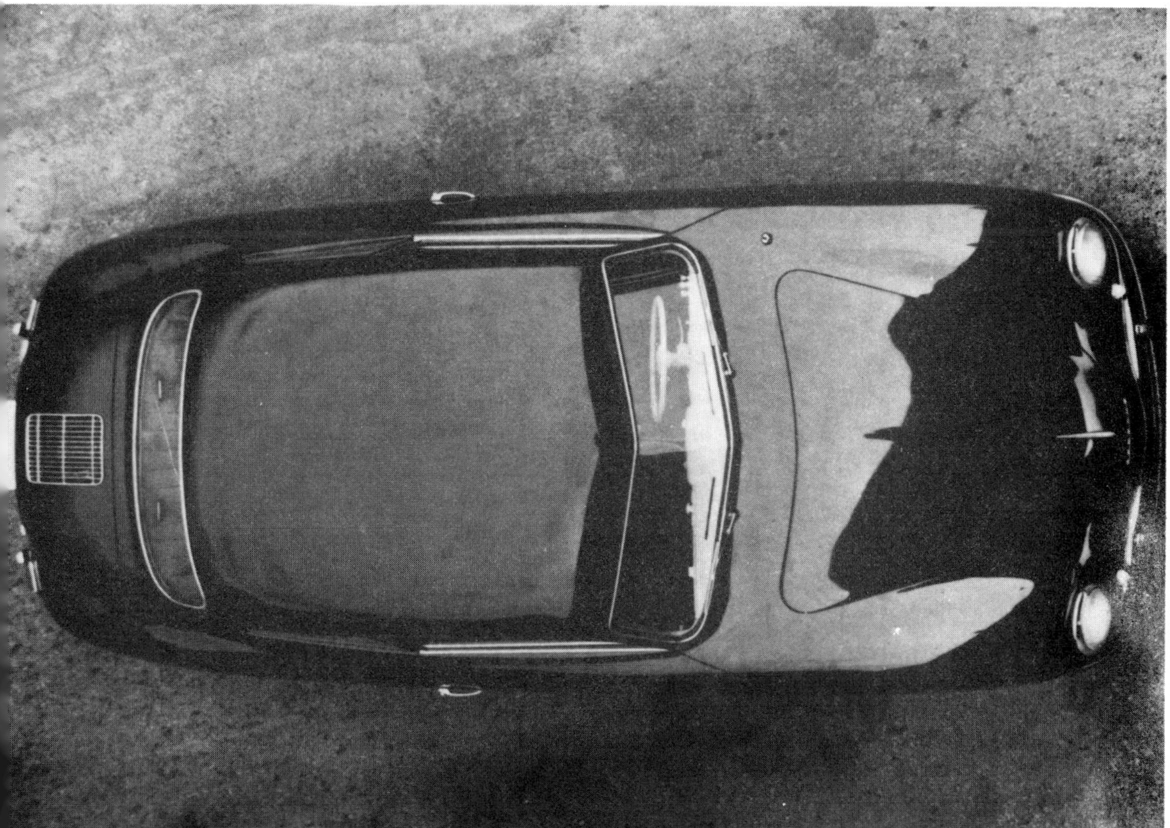

Porsche prototypes: Types 534/555/728

A tradition going back fifty years links the firms of Porsche and Volkswagen in close technical co-operation; after all, Professor Porsche designed the VW Beetle and, together with the town of Wolfsburg, laid the cornerstone for this car factory.

The development and contract work awarded by Volkswagen to Porsche would alone fill several books. Some projects were unimportant, others are still secret, so we will look at just some of the more significant contracts.

Even when the Porsche company was just becoming recognized, early in the 1950s, the Stuttgart designers presented the Wolfsburg firm with variations on the Beetle. In particular, the small sports car based on a VW platform bore a close similarity to the 356.

This Type 534, made in 1952/53, was a coupé on a shortened VW platform, the wheelbase being 82½ in (2,100 mm). An engine of just below 1,000 cc developed 26 bhp, with a compression ratio of 6.5:1. Porsche provided the four-speed gearbox, while the clutch and most of the mechanical parts came from VW's own bins. The aim was to produce a car that weighed no more than 650 kg (1,433 lb), and the front:rear weight distribution was 43:57. The prototype was presented in the autumn of 1953 to VW's then general manager, Heinz Nordhoff, but the project was not adopted and the prototypes were scrapped on VW's instructions.

Another typical order from Volkswagen to Porsche was the Type 728 (EA 53). Porsche had already made initial drawings for this car in 1954, but it was not until 1960 that the prototypes could be made. The front section was strongly reminiscent of the VW Type 3, with the rear designed either as a 'fastback' or as a 'notchback'.

But this project, also intended as a successor to the Beetle, never got beyond the trial stage. The engines were far too puny (0.8 to 0.9 litres, 26 to 32 bhp), and, although a 1.3-litre developing 42 bhp was quickly installed, the thumbs-down came from Wolfsburg about a year later.

Opposite page *A study by Porsche for a small VW sports car, Type 534.*

147

148

Opposite page *The Type 555, a four-seater Volkswagen with a rear-mounted engine.*

This page *Porsche development 728 with subtle differences (wheels, indicators, bumpers, badges and rear-end treatment).*

Porsche: Type 695

Even in 1959, when the Porsche 356 was still in its prime, a programme was approved for the design of a new, four-seater Porsche sports car based on this 356. At the end of that year Ferdinand Alexander 'Butzi' Porsche, head of design, and his team started work on the 695 project (or 'T7' as it was known internally; T5 was the Porsche 356 revision of 1959, and T6 the 356 revision of 1961).

Porsche took over the 94½ in (2,400 mm) wheelbase of the experimental Type 530 (described on page 144) and designed a body not very far removed from the 356 styling, although it also pointed to the later Porsche 901/911. The wing contours were even more prominent, the bumpers were integrated in the body shape, and large areas of glass were featured above the waistline. The size of the rear window, in particular, was impressive.

The interior included a new design of instruments and mountings, while, in the rear, semi-bucket seats could accommodate two adults—with a certain amount of goodwill! The backs of the rear seats could be tilted forward, offering considerably more luggage space.

By the winter of 1961 the Type 695 had reached the test-driving stage. The rear substructure was almost identical to the 356's, but at the front Porsche had installed a MacPherson strut suspension. There remained one major problem: the existing four-cylinder power units were either insufficiently powerful, or too expensive for series production. So, in 1961, an order was placed for the development of a six-cylinder engine (Type 745).

The 2-litre 'boxer' flat-six engine, cooled by two axial fans, was already being tested in 1962, and produced 120 bhp at 6,500 rpm. This was still deemed insufficient, so a 2.2-litre version was made, which gave the required 130 bhp.

The engine was ready for Type 695 early in 1962 . . . but changes in personnel also resulted in a new design strategy! Ferry Porsche finally decided against the four-seater car and reintroduced the original concept of a classic two-plus-two coupé.

One good thing had come out of the Type 695 project, for its detail drawings served as an inspiration for the '901' seen at the Frankfurt Show in 1963, which became known as the 911 a year later. The forerunner of this model as well as its engine—which also never went into production in this form—can be seen in the Porsche museum at Zuffenhausen.

The T7 prototype, or Type 695, was the forerunner of the Porsche 911. It was produced by the design studio during 1959-61.

In the winter of 1961 the Type 695, precursor of the 911, started its test programme.

Porsche: 911 four-seater

A four-seater Porsche 911, with the internal type number 911/C 20, was manufactured on July 6 1970. The body (chassis number 911.030.0004) had been extended by 11¾ in (300 mm) in front of the cross-member, while the front fixing points for the front suspension wishbones had been shifted upwards by about 1¾ in (45 mm), although a thick shim brought the wishbone back to its standard position. The upper shock absorber mountings points, and caster angles, were modified.

The wheelbase of this 911, which still had two doors, was extended from 89.3 in (2,268 mm) to 103 in (2,615 mm), although the turning circle turned out to be unwieldy, averaging 40 ft (12 m)! Unladen weight rose to 2,482 lb (1,126 kg), with a front/rear weight distribution of 43/57 per cent. Although this was not a final specification—as evidenced by a note at the experimental department in Weissach—the four-seater remained a one-off, considered to be rather better than Pininfarina's attempt to convert a 911 Coupé to a four-seater (see page 132).

A 2.2-litre six-cylinder engine developing 180 bhp was used in this experimental car. The five-speed gearbox was the earlier design, since the stronger Type 915 gearbox could have been fitted only after extra modifications to the body. The standard disc brakes were taken from the 911S, as were the 6 in wide Fuchs wheels and Michelin 185/70 VR 15 tyres. Koni dampers were fitted at the front and rear, as were 0.59 in (15 mm) diameter anti-roll bars. The rear axle had to be modified, since the trailing links were extended by 2 in (50 mm).

This four-seater 911, under its internal C20 reference, remained a one-off (1970).

Porsche: 914 eight-cylinder

The 'wolf in sheep's clothing' philosophy was almost perfected in this mid-engined Porsche. Externally the car was just like the original 914, apart from the additional engine oil cooler mounted below the front bumper. Inside the car, the only giveaway feature was a 10,000 rpm rev-counter!

The 908, eight-cylinder racing engine was the chosen power unit for the car, which was built for Ferdinand Piëch, the then head of development who had a taste for speed. In race tune the twin camshaft engine would produce a reliable 360 bhp, while with a road exhaust system it provided a good 300 bhp, with plenty of power and torque from 2,500 rpm upwards. Even in the early 1970s it cost an estimated DM 100,000 to carry out this conversion.

The sound of the engine running at over 7,000 rpm, and the eight Bosch injectors operating at full pressure, remained a lasting memory for bystanders. The lack of sound-deadening air filters was, however, a considerable annoyance to the occupants. This problem was alleviated to some extent in the second 914-8 to be built.

Dr Ferry Porsche received this car as a gift from the workforce on his sixtieth birthday. Equipped with four twin-choke carburettors, and air filters, the power was further reduced by some 30 bhp, but still had ample reserves. Dr Porsche's 914-8 conformed rather more to the concept of a production car, although for reasons of stiffness and safety the usually detachable roof was bolted firmly into position.

The high-quality finish was retained, and the wheel rim width of 7 in remained quite civilized. Dr Porsche drove his birthday present for over 6,000 miles (10,000 km) on open roads, with a potential maximum speed of more than 140 mph (230 km/h).

An eight-cylindered Porsche 914 was built for Dr Porsche. It was powered by the 3-litre 908 racing engine detuned to 270 bhp, and was capable of 143 mph (230 km/h). The original 914-8, built for development chief Ferdinand Piëch, was virtually a racing version.

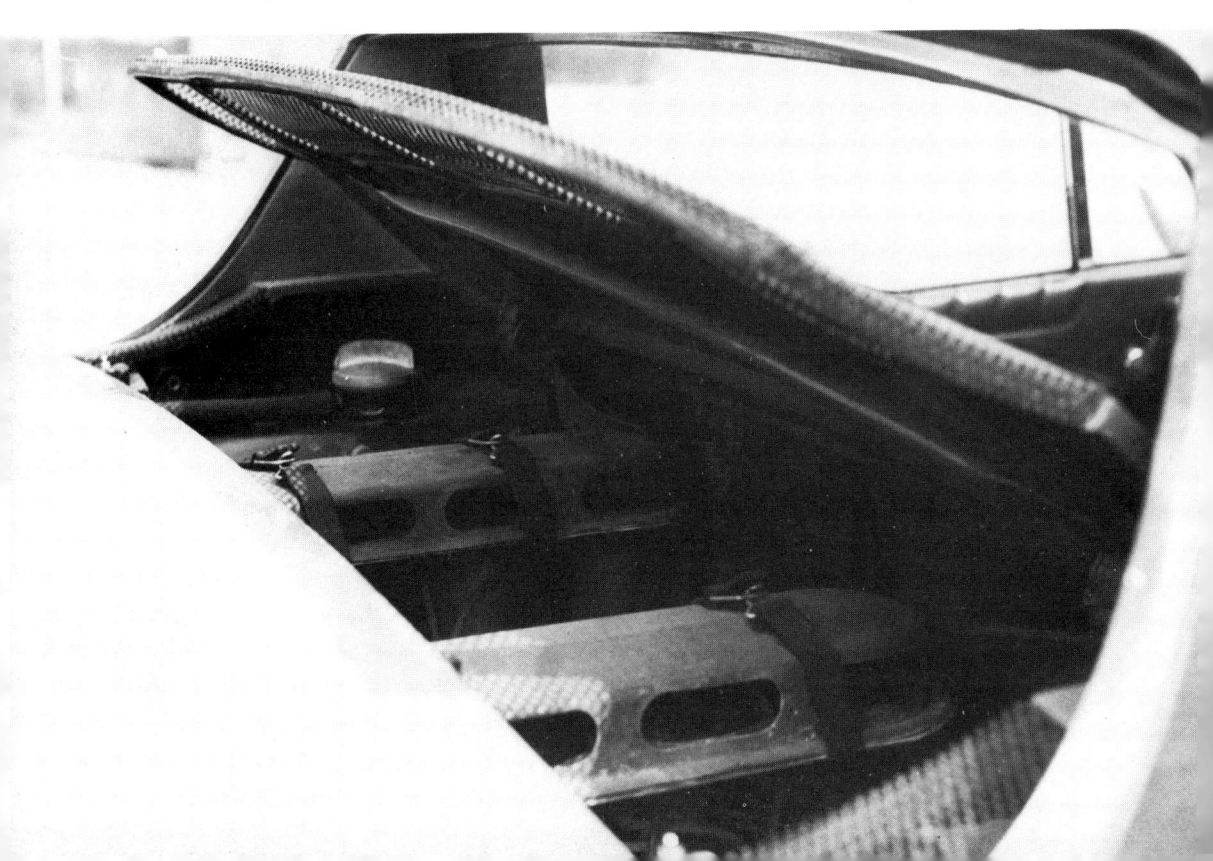

Porsche: 928 four-seater

Fifteen years later Dr Ferry Porsche received another special birthday present—a four-seater 928S. For his 75th birthday, celebrated on 19 September 1984, the workforce presented him with a special car which had been completed by the prototype department at Weissach in just nine months.

The 928's bodyshell was extended by 10 in (250 mm), and in order to make it easier for people to get into the back seats the B-pillar was made more upright. Then, to give them the maximum headroom, the roofline was kept as straight as possible, so that in side view the four-seat 928S looked like an extended hatchback.

At Dr Porsche's request, the front wings were raised on the 928-4, to provide 'sighting lines' for parking and traffic driving.

The power unit chosen for this car was the brand-new 5-litre V8 with four-valve cylinder heads, developed for the American market, though rated at 310 bhp. In a car weighing 3,580 lb (1,625 kg), the performance data included a maximum speed of around 160 mph (260 km/h), and acceleration to 100 km/h (62 mph) in 6.5 seconds.

The latest Daimler-Benz four-speed automatic transmission was fitted, and the chassis specification remained unchanged, apart from harder springs and shock absorbers at the rear.

Dr Porsche was particularly pleased with the interior, which boasted, among other features, computer-programmed front seat settings, a telephone, tempostat speed control, an alarm system, and a Blaupunkt stereo radio system.

Dr Ferry Porsche's 75th birthday present from the workforce was a genuine four-seater 928S.

Porsche-Allgaier: tractors, Types 110/111/113/425/535

Professor Porsche was able, in his lifetime, to see for himself the fruition of a plan laid in 1937 to produce tractors. The 40th German Agricultural Show in Frankfurt, in 1950, saw the debut of 'his' tractor, Type AP 17.

Many features of the Porsche tractor were new and advanced: the air-cooled engine had a comparatively high running speed for that time, featuring lightweight construction for the engine and transmission. The centrifugal oil filter, the porch-shaped axle, track width adjustment, the hydraulic clutch, and not least, the model's competitive price, were all in its favour.

The 'Porsche Diesel Engine construction' firm was based in Manzell, a suburb of Friedrichshafen on Lake Constance. Within six years four different tractor designs were offered: an 11 bhp single-cylinder, a two-cylinder with 22 bhp, a three-cylinder with 33 bhp and, logically, a four-cylinder with 44 bhp.

A common bore/stroke ratio of 92 mm × 116 mm was retained and, in effect, the larger capacities were achieved by adding cylinders, which was handy for parts storage and repairs. The capacity of each cylinder was 822 cc.

Air cooling was provided by a fan, while a remote thermometer and an audible warning signal alerted the operator when the maximum cylinder temperature was exceeded. The track width could be adjusted for various requirements—on the smallest tractor, for instance, from 39 to 65 in (1–1.65 m)—and all versions had a very small turning circle. It was possible to brake the rear wheels individually, and turn the tractor sharply with one wheel locked.

Below the four- or five-speed gearbox was a special gear enabling the operator to drive at just 770 yd (700 m) per hour. The tractor could be started in any gear, since the hydraulic clutch made it impossible to stall the engine. In fact the engine did not cut out but was declutched, the engine continuing to run at tickover speed when the brakes were applied.

Introduced mainly as standard features (although some were optional), were hydraulic power lifts and/or various power take-off shafts for driving attachment machinery, and other agricultural equipment. Even ten years after the launch of the first Porsche tractor, the same gearbox as in the AP 17 was still used in the 'Standard' and 'Super' models of the Porsche diesel tractor. Over 50,000 gearboxes had been cast from the original mould, as well as at least 150,000 identical cylinder heads. Even at this stage, a 22 bhp replacement engine, or even a 25 bhp unit from the later series, would still fit the AP 17 ancestor, because flange sizes and dimensions of many individual parts remained unchanged over the years.

At the end of the 1950s the Junior was the smallest Porsche tractor, its single-cylinder engine producing 14 bhp, and its fuel consumption averaging 0.2 gal (1 litre) per hour over a year. The Standard was a direct descendant of the AP 17, its two-cylinder engine producing 25 bhp. The Super three-cylinder diesel tractor produced 36 bhp, and for all models there was the option of the hydraulic clutch, under the designation 'H'.

The agricultural tractor (Type 110, model 1), appeared on the German market in about 1939.

In the S7 version the driver sat at the back with a loading platform at the front.

The S2 version had rear-mounted driver's seat and steering gear.

The 110, fully equipped with all the agricultural accessories.

158

Porsche designed the new tractor series as early as 1948 (Type 425/111-13).

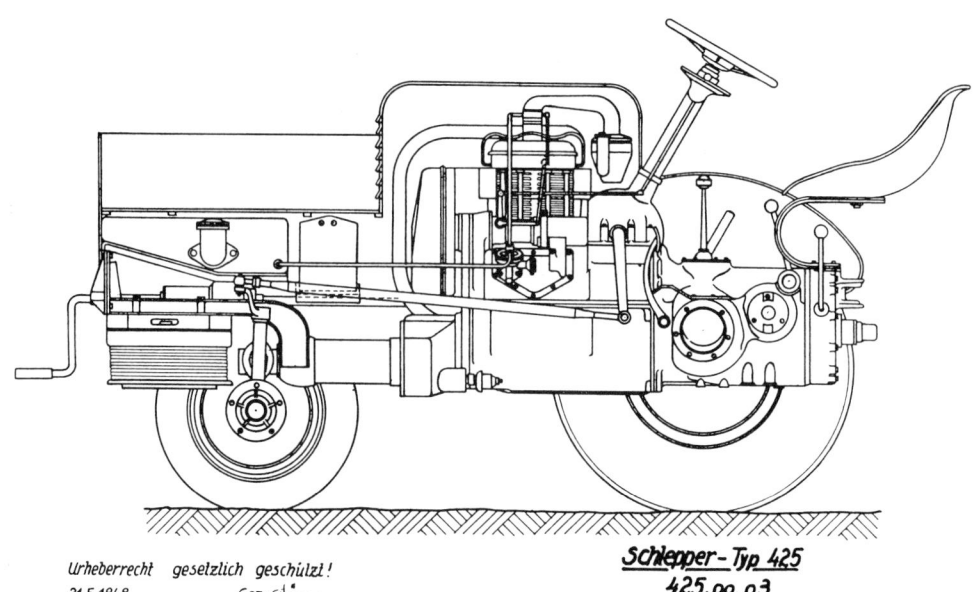

Urheberrecht gesetzlich geschützt!
21.5.1948   Gez.:

Schlepper-Typ 425
425.00.03
PORSCHE-Konstruktionen Ges.m.b.H.
Gmünd, Kärnten

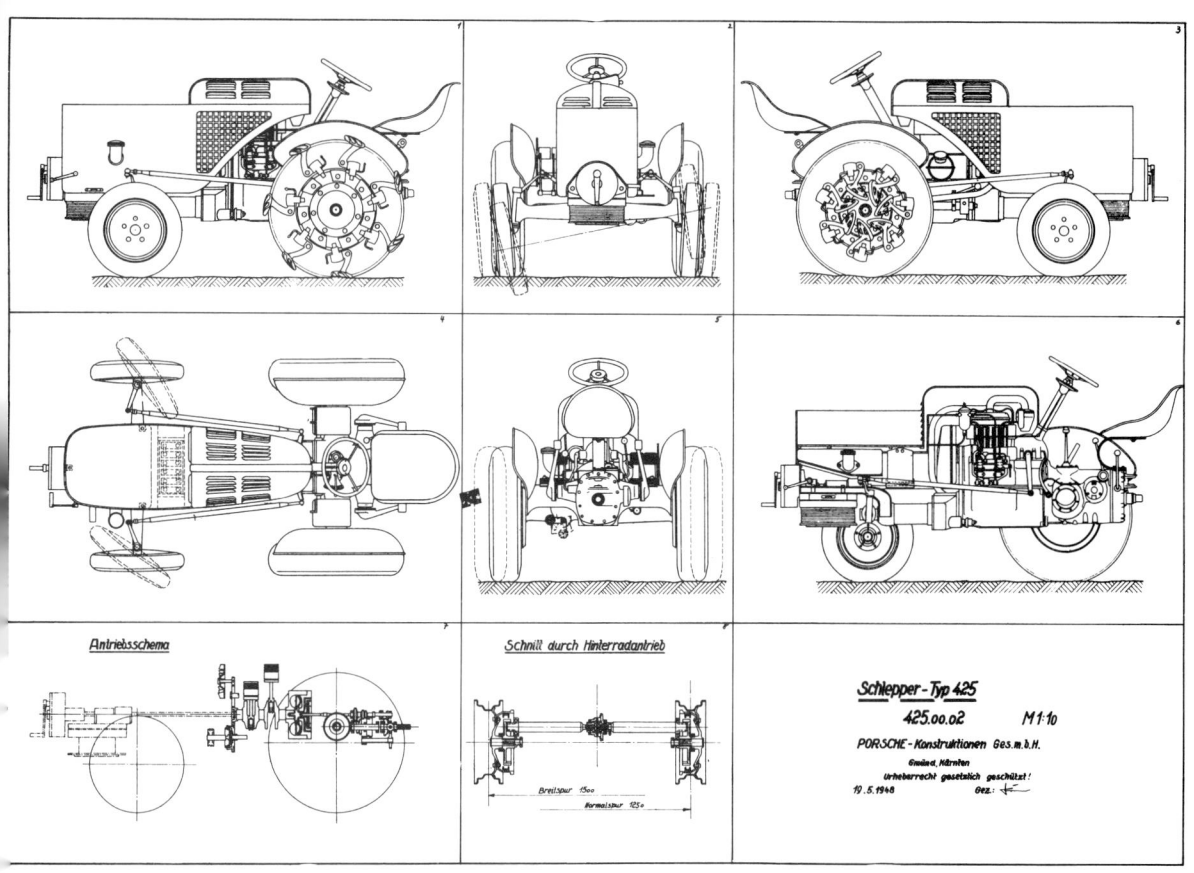

Drawings for the tractor Type 425—later known as 111 to 113.

The diesel engine, with from one to four cylinders, provided between 11 and 44 bhp.

A special design for plantation use, the Type 535 of 1953/54 was based on the 111.

Porsche-diesel (Allgaier) in the A 111 version (12 bhp), built in about 1954.

Porsche-diesel: Types 2086-2089

The diesel engines fitted in the Allgaier tractors (described in the previous entry) are Porsche developments. From Type number 2086 there was a series of fast-running one- two- three- and four-cylinder air-cooled engines of 12 to 48 bhp. They combined a compact design with considerable standardization of components, which made for inexpensive production and low-cost maintenance.

Because of their good power-to-weight ratio, the new diesel engines were suitable both as industrial engines for generator, ventilator, pump and compressor units, and for tractors, agricultural machinery and similar applications.

The in-line, air-cooled cylinder assemblies can be replaced individually, and the layout facilitates maintenance work. A space-saving, high-efficiency cooling fan is mechanically attached to the engine, dispensing of course with water and circulation systems, so it can be operated efficiently in all climatic conditions.

Detailed data
Single-cylinder (Type 2086) Bore/stroke 88 × 96 mm, capacity 584 cc, compression ratio 20:1, power 12 bhp at 3,000 rpm, swirl-chamber injection with pintle nozzle and rod glow plug. Dry weight 330 lb (150 kg).
Two-cylinder (Type 2087) Data as Type 2086, but with 1,168 cc capacity, 24 bhp at 3,000 rpm, and 395 lb (180 kg) dry weight.
Three-cylinder (Type 2088) Data as Type 2086 but 1,752 cc capacity, 36 bhp at 3,000 rpm, and 463 lb (210 kg) dry weight.
Four-cylinder (Type 2089) Data as Type 2086, but 2,336 cc capacity, 48 bhp at 3,000 rpm and 533 lb (242 kg) dry weight.

The three-cylinder diesel (Type 2088) for the Allgaier tractor developed 36 bhp.

Type 2089: four-cylinder diesel engine with 2,336 cc displacement giving 46 bhp at 3,000 rpm.

Research car: Type 995

The Porsche 995 is a futuristic research car developed at Weissach, in response to a contract awarded by the German Federal Ministry for Research and Technology (BMFT). The brief was to design a car that would be light, strong, safe, have good performance and economy, yet be 'realistic' and saleable.

When it was unveiled, in the middle of 1979, the 995 proved to have an average fuel consumption of only 34 mpg (8.3 litres per 100 km). This could be achieved without any reduction in performance by optimizing the engine and the power transmission, and through careful attention to aerodynamics and weight.

The research for the sports car of the future produced the concept of a compact, streamlined vehicle with an aluminium body, and a drag coefficient of only 0.30.

The design proposed two alternative engines which, after taking all considerations into account, both burnt petrol rather than diesel fuel. The first option was a four-valve, four-cylinder, 2.2-litre engine with very favourable exhaust emission values. The second option was a V8 engine with a capacity of 3 litres. In order to allow these engines to run at maximum efficiency, an automatic gearbox with two mechanical clutches, allowing gear changes to be performed under load without interruption of the drive, proved to be ideal.

Design study for a sports car of the future, using new technology.

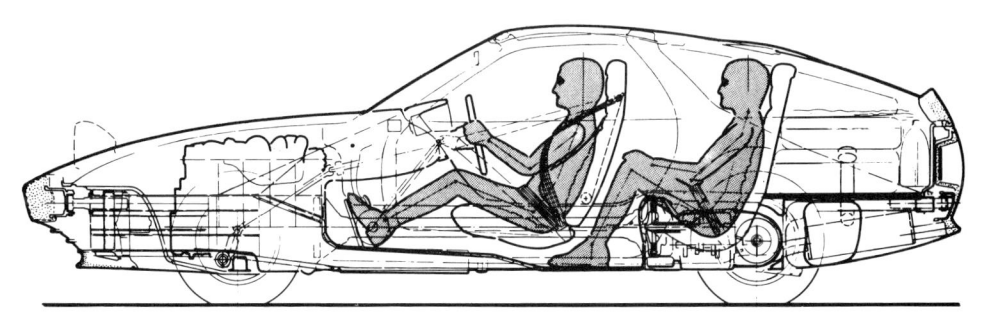

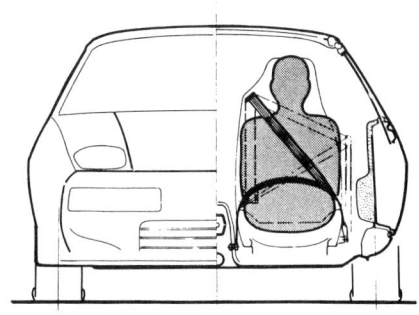

Rossi: 917 base

Count Gregorio Rossi di Montelera, head of the Martini & Rossi drinks company that had sponsored the Porsche factory team for some years, caused some headaches when he said he wanted a 917 for use on the public roads. Some problems had to be overcome, since Count Rossi wanted the car to be as original as possible, and only allowed a silencer to be fitted, along with a safety guard over the cooling fan, to keep out stones.

In 1975 one of the last 917K racing cars in Porsche's possession was handed over to Count Rossi. Because it was a German racing car it had to be sprayed silver, on the same principle as other cars in his collection which were painted French blue, Italian red and British green. A certificate of roadworthiness was obtained in the State of Alabama!

A trial-run two months before the handover went well, Count Rossi expertly driving the 620 bhp coupé around the Weissach test track.

At the handover, on 28 April 1975, Count Rossi received a manual containing a complete history of the car. It had been built in January 1971 with the chassis number 917.030 and raced once, with the start number 28, in the hands of Helmuth Marko/Gérard Larrousse—in Martini & Rossi colours, of course. This particular 917 was unique in that it was the first Porsche to be raced with anti-lock braking. The race had not gone well for 917.030, but it was later used for further development on anti-lock braking, then mothballed until it was prepared for Count Rossi's personal use.

When he collected the car, Count Rossi, accompanied by his personal secretary, left Stuttgart at four o'clock in the afternoon, and was safely in Paris before midnight. No problems had been encountered with the 917, which proved to have an average fuel consumption of 9.41 mpg (30 litres per 100 km) on 4-star petrol.

The only delay occurred at the German-French border, when the customs official called over a gendarme to help classify the Porsche. The gendarme pronounced the 917 perfectly acceptable for use on the French roads . . .

A roadworthy Porsche 917 was prepared by the factory for Count Rossi. It had been built in January 1971 and was raced once, in the Österreichring 1,000 km.

After the Austrian race, in which the 917 ran with anti-lock brakes, the special became a test car, before being fettled for road use in April 1975.

Sauter: 356 base

Shortly after the Second World War, racing and rallying started up again in Germany, although at a much lower level than the exalted Auto Unions and Mercedes of the pre-war times. In the early 1950s enthusiasts were looking for suitable cars, and Porsche provided them. Some did not regard the 356 Coupé as ideal for every type of event, so different versions were to be seen left, right and centre.

Heinrich Sauter, an industrialist from Stuttgart, was one who had his own ideas, building himself a notably lightweight roadster body on a standard 356 chassis. The work was carried out in the workshop of Hans Klenk: the result weighed less than 1,325 lb (600 kg), and was powered by a Porsche 1500 engine. Mechanically, it differed only in having special spring-hub wheels, designed by the engineer Mantzel.

The Sauter-Porsche was not very successful, and at the end of 1951 it was sold to a Frenchman, François Picard, who sprayed it blue and continued to compete, naming the car 'le petit tank'.

SAVE: ambulance study Type 2539

One of Porsche's more unusual undertakings, in the mid-'70s, was to design and produce a prototype for an ambulance of the future. The contract was awarded by the German Federal Ministry for Research and Technology, which gave Porsche AG the brief to define the requirements for medical and technical equipment based on analyses of ambulances already in use.

Specifically, existing vehicle components, or those under development in the motor industry, were to be employed to ensure that the eventual design would be thoroughly practical. The end result was a system called SAVE, the German acronym for 'Swift ambulance first-aid'.

The project manager decided that the first prototype would be based on the Volkswagen LT and the Daimler-Benz 207, both vehicles having a 115 bhp engine. The ambulance unit needed to be well insulated from road shock, of course, and Porsche's prototype featured a mechanically interconnected suspension between the chassis and the ambulance body. Suspension and damping were by four standard Porsche struts. All the main components, such as cross-members and suspension arms, were cast in aluminium.

The ambulance body, made of sandwich-construction plastic panels, was made in two parts, upper and lower, and could be lowered on to the chassis by means of four struts, hydraulic or mechanical, which could be carried on the vehicle during the journey. The stretcher table was positioned so that the patient's head would be in a position least vulnerable to vibrations, and medical equipment was stowed in handy boxes. Air conditioning, ECG unit and oxygen apparatus were integral to the design.

Basic and additional features were as follows: good travelling comfort, and ergonomical design to reduce movement between the doctor, the patient and the ambulanceman; plastic construction for safety; good noise and temperature insulation; supply chutes incorporated in the base unit; easy disinfection through integral manufacture.

Units required only for outside use were accessible only from the outside; there were connections for external oxygen and energy supply; the safety of attendant personnel was guaranteed; and additional stretchers could be fitted with little loss of space. Two intercom systems (in the driver's cab, and in the ambulance unit) ensured easy communication, and there was maximum portability of the internal equipment.

The SAVE ambulance system (Type 2539), designed and built to a contract from the German Research Ministry. SAVE was an acronym for 'swift ambulance first aid' in German.

The base vehicles are the Volkswagen LT or Daimler 207 (illustrated).

The ambulance unit is supported on four hydraulic struts.

Sbarro: 911 Turbo engine

Franco Sbarro, the Swiss specialist in customized cars, is renowned for his unusual designs. His workshop at Lake Neuchâtel has produced many 'freaks' which amaze visitors to motor shows, often with modern components under plastic replicas of historic cars.

One of his more unusual creations was a Volkswagen Golf into which the former teacher inserted a Porsche 911 Turbo engine. The 300 bhp six-cylinder unit took the place of the rear seats and drove the rear wheels.

As though this were not enough, Sbarro arranged for the owner to be able to show off this arrangement to his friends: he devised a mechanical strut system which, at the touch of a button, raised the back of the Golf high into the air so that the Porsche motor could be seen (and serviced) from all angles.

A most unusual Volkswagen Golf . . . Franco Sbarro's 'folding' version with a 300 bhp Porsche 911 turbo engine—and the means to show off the hardware!

Silver Satin: 356 base

Dean Jeffries, a Californian body specialist, bought himself a Porsche 356 Coupé at the end of the 1950s and set to work to modify it to his own design and standards. To start with he stripped the car down to component form, inside and out, and removed all the paint. His objective was to build a super-Porsche without losing the typical appearance of the marque.

Jeffries threw away the bumpers, and made new front and rear bodywork out of aluminium. Particularly at the rear, the car bore a resemblance to the popular Mercedes 300 SL. The interior was completely restored by this individualist, and the car was painted silver, with items like the engine, wheel rims and tool kit either silvered or chromed.

The engine, a standard 1.6-litre Carrera, nominally producing 105 bhp, was uprated by about 20 bhp using the factory's racing sports kit. The whole job took Jeffries seven months to complete, in 1959, and cost some $8,000.

The Californian body specialist, Dean Jeffries, reshaped his 356 with attractive front and rear body sections, made in aluminium.

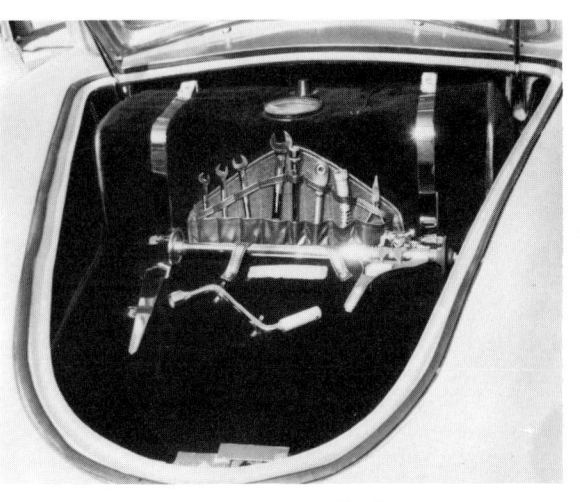

Above left *The luggage compartment of 'Silver Satin' contained a complete, chromed toolkit!*

Above right *With a factory sports kit fitted, the 1.6-litre Carrera engine had its output increased to 125 bhp.*

Simpson-Chevy: 911 base

In America, where nothing seems to be impossible, there are people who believe that Porsche production models should go even faster. For those who subscribe to the theory, designer Rod Simpson from Santa Monica, California, offered the brutal solution: a Porsche 911 powered by a Chevrolet V8 engine!

Transplanting the Chevrolet V8 into the rear of a 911 involves a lot of modifications, and bracing too. Voluptuous wings are flared to cover the wheels, 235 × 14 in at the front and 285 × 15 in at the rear. The braking system remains unchanged, but the car's interior is radically altered, since the V8 is placed where the back seats should be. A full roll-over cage is fitted inside the car. Standard drive-shafts are retained for the 200 bhp vehicle, but racing starts are not recommended since the engine produces so much torque.

The rear spoiler is much larger and the extra weight of the American engine is compensated, to some extent, by the omission of interior fittings and equipment, and by the adoption of plastic materials for items such as the enlarged front lid. A radiator is fitted at the front to provide cooling for the V8, and has the advantage of ducting hot air over the windscreen . . . Simpson claims that this protects the glass from stone damage!

Several 'Porschev' 911-Chevrolets were made and sold, Simpson offering the parts in kit form for home assembly.

Steel Special: 911 Coupé

A special Porsche was handed over to the German Museum in Munich at a small ceremony in October 1974: a 911S with a bare metal body made of stainless steel.

The gleaming, silver-coloured 911S had reached the end of its proving trial after seven years of hard driving, which began straight after the car's debut at the International Motor Show in Frankfurt in 1967. Heinz Todtmann, director of the German Stainless Alloy Steel Information Office, had driven this otherwise completely standard Porsche both in Germany and abroad throughout the seven years. The stainless-steel Porsche was just a prototype, to demonstrate the reliability of the material, and there was never any plan to put it into regular production. Three were made, but only one survives.

Porsche 911 with a special stainless-steel body. It was tested over 150,000 km (93,205 miles) between 1967 and 1974.

Storez: Carrera GT base

The French GT Champion Claude Storez went to the Zagato factory near Milan early in 1959 to have a very light, attractive body fitted to his 356 Carrera GT. The car was taken to the Porsche factory for final adjustments, and was then to be driven to Paris by a French mechanic.

Unfortunately, the journey ended a few miles north of Stuttgart, where the mechanic had a bad accident which destroyed the Storez-Porsche. The driver was not hurt, but the car was written off.

The attractive Zagato lightweight body for the 356 Carrera. Claude Storez, the 1958 French GT champion, realized his dream for only a very short time, as the car was written off in a traffic accident as soon as the work was completed.

Studebaker: Type 542

The first contact between Porsche and the American manufacturer, Studebaker, came in December 1951. Dr Ferry Porsche made his first post-war visit to America to discuss a military vehicle concept, and to make further plans with his New York importer of the time, Max Hoffmann.

Hoffmann had convinced Studebaker's export vice president, Richard A. Hutchinson, that America really needed a domestic 'Volkswagen', suggesting that Porsche would design it, and Studebaker build it. Hutchinson, meanwhile, succeeded in obtaining a contract from VW-Wolfsburg for Studebaker to be the US licensees for the VW marque . . . only for the head of Studebaker to turn the offer down! That just left the original proposal on the table—for Porsche to design a popular car to be produced by Studebaker.

In May 1952 a Porsche delegation including Dr Ferry Porsche, Karl Rabe, Leopold Schmid and Erwin Komenda, made another visit to Studebaker's headquarters in North Indiana. They took with them a prototype of the Type 530, which was a Porsche 356 with a longer chassis and seating for four people, though with only two doors.

The day after their arrival a trial-run took place with Hoffmann, Hutchinson and Studebaker president Vance. It was a dreadful demonstration, as Hoffmann remembered it, since the lengthened 356 was extremely noisy when driven on rough roads. Nevertheless a contract was signed for the construction of a prototype featuring a more powerful engine, to be sited at the front, and weighing less than the Studebaker Champion, which was quite successful at the time. The engine was to be a six-cylinder, with air cooling, and the car was to have three forward gears and a top speed of 84 mph (135 km/h).

The Porsche team started on the design as soon as they returned to Stuttgart, starting with a completely new 120° angle V6 power unit. The engine had a mixed cooling system with an air-enveloped cylinder head, and blower cooling for the cylinders.

By August 1952 Ferry Porsche and Karl Rabe had a 1:5 scale model to show to the Americans. After lengthy discussions about the mixed cooling system, it was agreed to build a water-cooled six and an air-cooled version, and to test them both. Once a new contract had been signed, the Studebaker project went into full swing, early in 1953, under the project number 542.

By the end of 1953 a pontoon-shaped four-door body was ready. It was lower and shorter than the '52 Champion, but incorporated some technical details of the American car, notably the door fastenings, steering gear and steering wheel, three-speed automatic transmission with overdrive (from the Commander), as well as brakes and tyres.

The 542 L-engine (air-cooled) and 542 W-engine (water-cooled) were very similar in their basic design, each having a bore and stroke of 90×80 mm, and a capacity of 3,054 cc. The air-cooled engine weighed 483 lb (219 kg) and achieved 98 bhp at 3,700 rpm, while, unexpectedly, the water-cooled engine was lighter at 454 lb (206 kg) and developed 108 bhp at 3,500 rpm.

Two representatives from Studebaker drove the 542 for the first time in March 1954, on the way back from the Geneva Motor Show which Dr Porsche attended with the car. After a final trial in August 1954, the 542 was taken to America.

Even though the Porsche design had become a good 550 lb (250 kg) heavier than the Studebaker Champion, and the air-cooled engine offered lower performance than the water-cooled unit, the people at Studebaker were astonished at how much had been achieved in such a short time.

But then came one setback after another. Just a few months after the 542's presentation, Packard took over the financially embarrassed Studebaker company, and the capital of 15 to 20 million dollars to construct the V6 engine was simply not available.

At the end of 1954 co-operation ceased between Porsche and Studebaker. Only the water-cooled engine remains now, on view in the Porsche museum in Stuttgart. Dr Ferry Porsche continued for a while to drive the Studebaker Commander, a really successful two-door coupé produced by the designer Raymond Loewy.

Below *Contracts for the construction of Studebaker prototypes, designed by Porsche, were signed in 1952.*

Opposite page *The Type 542 was ready by the end of 1953. It had a pontoon body and was powered by a new, Porsche-designed six-cylinder engine of 3,054 cc capacity, developing 98 bhp with air cooling or 108 bhp with water cooling. The project was terminated when Studebaker got into financial difficulties.*

177

TAG-Turbo: Formula 1 engine

Porsche designed and developed the 'TAG Turbo PO1' engine, on behalf of the TAG Turbo Engines company based in Geneva, with outstanding success. The engine made its debut in the Dutch Grand Prix at Zandvoort on August 28 1983, in the Marlboro McLaren International's MP4/1E car, driven by Niki Lauda.

This TAG engine will go down in history for its rapid development and complete success. The development contract was signed at Weissach on 12 October 1981, and straight away the designers began work from scratch. By 18 December 1982 the first development engine ran on the test bed, and the first driving trials took place in March 1983. The car ran in its first race only twenty weeks later.

In the last race of the 1983 season, in South Africa, Niki Lauda was contesting second place when a minor defect, not related to the engine design, put him out. The signs were clear for the opposition, and throughout the 1984 season Lauda and his team-mate Alain Prost completely dominated the Grand Prix scene, winning 12 of the 16 races. Niki Lauda won the drivers' championship closely followed by Alain Prost, and the McLaren-TAG team won the constructors' championship convincingly.

The 1985 season was another success for the TAG engine, which helped Alain Prost to win his first drivers' championship, and the McLaren team to carry off the constructors' championship for the second year running.

The outstanding features of the TAG Turbo PO1 engine, an 80° V6, are its reliability, power, compact design and low weight. Important, too, is the good fuel economy which has enabled drivers to cross the finishing line at racing speed, whilst rivals have slowed or dropped out because their engines have run out of fuel.

The TAG engine's output is in the region of 800 bhp at 11,500 rpm. For Porsche, the Formula 1 engine is but one of many outside development projects under way at the Weissach development centre.

Specification
TAG Turbo engine, TTE-PO1
V6 engine with twin turbochargers and intercoolers: *V-angle* 80°; *Bore* 82 mm; *Stroke* 47.3 mm; *Capacity* 1,499 cc; four valves per cylinder; four camshafts, gear driven; one-piece aluminium alloy cylinder heads; aluminium/magnesium alloy crankcase; titanium connecting rods; water cooling, twin water pumps; *Total weight* 330 lb (150 kg); electronic ignition and fuel injection; water and oil pumps are located at the front of the engine, and are directly driven by the camshaft drive gears; *Engine power* 800 bhp at 11,500 rpm.

The TAG company awarded Porsche the contract for the design of a Formula 1 engine in October 1981. The V6 turbo unit made its debut at the Dutch Grand Prix in August 1983.

Installed in the McLaren chassis, the TAG-Porsche engine went from victory to victory through the 1984 and '85 seasons.

Designer Hans Mezger with Niki Lauda. Although the engine is rated at 650 bhp in race trim, from 700 to 900 bhp is available in short bursts.

Tapiro: 914-6 base

The leading Italian designer, Giorgetto Giugiaro, aroused excitement at the 1970 Turin Motor Show with his 'Tapiro' model, based on the Porsche 914-6. Giugiaro's Ital Design Studio completely redesigned the 914's body, turning the rather boxy mid-engined two-seater into a sleek, aerodynamic coupé with its doors hinged at the top.

Giugiaro extended the wedge-shaped nose in an almost continuous line with the windscreen. The gull-wing doors opened like those of the Mercedes 300 SL, and behind the cabin, divided by a strong central tunnel, was a narrow 'Targa' bar. Extending to the rear were two large cowlings, top-hinged like the doors, giving access to the engine and to the minuscule luggage compartment.

Under the conventional, forward-opening bonnet lid was the spare wheel, which incidentally filled the entire space. The long, narrow, wedge-shaped nose resembled that of the long-nosed animal called the tapir, hence the name, and it enclosed the headlamps which could be raised electrically at night.

To enhance its appearance the Tapiro-Porsche had 8×15 in rims at the front, and 10×15 in rims at the rear. The interior and instrument panel were to a completely fresh design, with the speedometer located in the centre, behind the twin-spoke steering wheel, and additional instruments on either side of it.

This mid-engined study, which never got beyond the prototype stage, incorporated a six-cylinder Porsche engine. The Italian tuner and racing driver Ennio Bonomelli had tuned the 2.4-litre engine to a good 220 bhp at 7,200 rpm, with a compression ratio of 10:1—theoretically good enough for a top speed of 143 mph (230 km/h).

Giugiaro presented the 914-6 'Tapiro' at the 1970 Turin Motor Show.

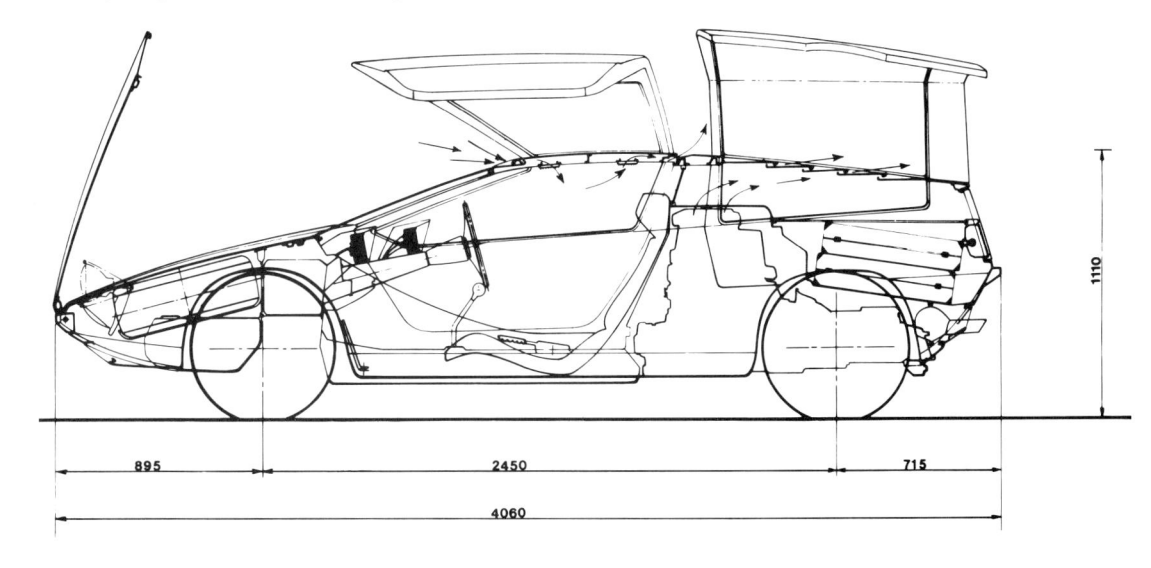

The 'Tapiro' on the Ital Design stand at the 1970 Turin Show had stylish gull-wing doors, matched by similar flaps for the engine bay. It was based on the 914-6 chassis.

Through-flow water turbines: Types 291/292/285

The through-flow water turbine (Type 291) was intended mainly for use with small water-power sources. It was therefore suitable for driving flour mills, saw mills, agricultural machinery and for generating electricity.

Compared with the water-wheel, the Type 291 had the advantage of a considerably higher speed, one of the benefits being that less power was wasted. Its high efficiency was maintained even with a considerable reduction in the flow of water.

There were two variants of the water turbine, 291/1 and 291/2, which covered a power range of from 4 to 40 bhp. For a gradient of, say, 1 in 5, the dividing line between the two designs was 11 bhp, at about 160 rpm. The turbines weighed between 530 and 650 lb (240 and 295 kg).

The turbine was constructed mainly of welded steel-plate sections and the runner, also welded, ran on ball bearings. The only maintenance required was the daily tightening and greasing of the bearing caps.

The Type 292 (1-3) through-flow water turbine was a more powerful version which delivered between 4.5 and 65 bhp, and this weighed between 990 and 1,875 lb (450 and 850 kg).

The water turbine (Type 291), designed for use with small water-power sources.

Townes: 911 base

When the last Porsche Speedster left the Zuffenhausen works at the end of 1958, a large group of enthusiasts waited in vain for a successor. Many of these were in America—particularly in California—which is where most of the $3,000 Speedsters ended up between 1954 and 1958.

Stan Townes was one of the 1958 356 Speedster customers, and he soon started to visualize what a successor model could look like: it would need to have more power, a five-speed transmission and disc brakes.

While Townes was completing his military service (he was demobbed in 1966) Porsche introduced the 911 model which had the necessary mechanical specification, but not the Speedster-style body. Townes became employed as a body builder with the Porsche-VW dealer in Santa Clara, California, and he waited his chance to buy a damaged 911, which presented itself late in 1967 in the shape of an overturned coupé with good mechanical parts.

Townes cut off the roof and then attended to the detail work of creating a speedster, with a new bonnet and tail, a raked windscreen and a slight widening of the wings. The work was complicated and laborious, so much so that Townes had to fabricate some parts several times over.

The task took three years to complete and cost around $15,000, but the speedster, painted metallic silver, was beautifully made and won many concours events.

The Weber-carburated six-cylinder engine was not tuned in any way, but was cleaned up and partly chromed—the exhaust pipes most noticeably. Goodyear tyres, 5.00/920-15 and 5.25/950-15, were mounted on the 7 and 8.5 in American-made rims.

Stan Townes' '911 Speedster' was built from a crashed coupé over a three-year period in the late 1960s.

New front and rear lids, and wider arches, were constructed by Stan Townes. There was also a speedster roof, not shown. The wheel rims were increased to 7 and 8.5 in wide, and were an American design. The neat tail treatment, improved by the deletion of bumpers, was enhanced by chromework. The three-year conversion was estimated to have cost $15,000.

VW mid-engined saloon: Type 1966

Before the Second World War Porsche's design studio created the 'Volkswagen', which later contributed to the rapid motorization of the German public. Since that time, and due to this historic connection, Porsche has been concerned to a greater or lesser extent with development work for VW in Wolfsburg. Two more cars which later went into production were the VW-Porsche 914, and the Porsche 924.

The 924 was originally intended as a sports car for VW, designed by Porsche. When VW had a change of management and a change of heart, Porsche bought back the design and offered it on the market under their own name. More than 130,000 Porsche 924s were made between 1975 and 1985.

Another vehicle was making its presence felt in the early 1970s, although only a handful of people saw it before it was literally crushed: the VW mid-engined small family car with the Porsche type number 1966, and VW's own designation EA 266.

Dr Kurt Lotz, who succeeded Heinrich Nordhoff as the chairman of Volkswagen AG in 1968, realized that the company would soon run into difficulties with a model range based on the Beetle. He therefore issued contracts to Auto Union, Porsche and NSU for a successor to the Beetle, but NSU soon pulled out, leaving just two companies working on the project.

The Stuttgart firm took an entirely new direction. Using the philosophy of the mid-engined VW-Porsche 914 design, Porsche gave the new saloon a mid-engine. This was a water-cooled unit, with four cylinders, located at the rear of a two-door hatchback reminiscent of the Polo model. The engine was practically under the rear bench seat—this being a full four-seater design—and the cooling air was ducted through slots in the body under the rear side windows, partly hidden by the wheel arches.

Using engine sizes of between 800 cc and 1,800 cc, the aim was to achieve 60 to 105 bhp, which would have given the most powerful version of this compact model a top speed of around 125 mph (200 km/h). However, as the engine needed to be fully enclosed to contain the noise and heat, it called for considerable expertise and would be very expensive to produce: even the base model, with the cheapest fittings, could not be sold for less than DM 10,000. In 1973/74, at the time of the intended introduction of the VW mid-engined EA 266, the VW Beetle cost little more than DM 6,000!

Although the engineers at Porsche, under the direction of Ferdinand Piëch, were also working on a two-seater version, as well as a two-plus-two coupé, and had almost completed their work, the end came quickly and unexpectedly.

Rudolf Leiding succeeded Lotz in October 1971, and found the company in bad shape. It had lost a considerable share of its market and the managers were restless, blaming Lotz for the unsuccessful development of the EA 266. In the meantime, Ludwig Kraus at Auto Union had successfully brought the Audi 100 on to the market and was working on the Audi 80 model. Both cars looked a better bet than the technically advanced, expensive EA 266.

Within a few weeks of taking office Leiding called a halt to the five-year

development of EA 266 and Porsche was instructed to scrap the prototypes
... most, but not all of them, ended up under the tracks of Leopard tanks!
VW took over the design of the Audi 80 from Kraus, changed the notchback to
a fastback, and introduced the model as the Passat. As for Piëch, he soon left
the Porsche company and joined Audi as engineering director, leaving behind
several prototypes which Porsche put away in a cellar. Maybe, like good
wines, they will be brought out one day to be savoured.

The small saloon for Volkswagen had a mid-positioned engine, and was type-numbered 1966 by Porsche, EA 266 by Volkswagen. After a five-year development period, the project was halted by incoming VW chief Leiding.

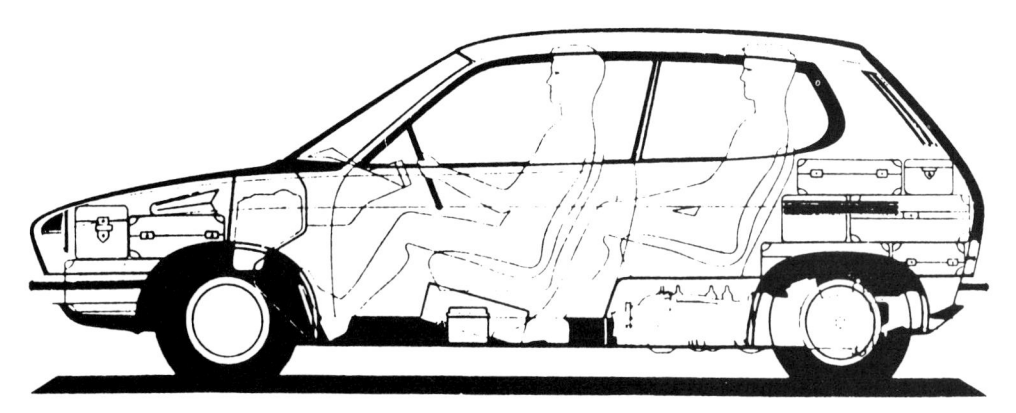

Volkswagen (Beetle): Type 60

The 17th of January, 1934, is significant in that it was the day that Dr Ing Ferdinand Porsche presented his 'Facts concerning the construction of a German people's car' to leading representatives of the Reich Ministry of Transport. Only a few weeks later, on 8 March, the idea was officially introduced at the opening of the International Motor and Motor Cycle Show in Berlin.

A contract was signed on 22 June by the Reich Institute of Motor Vehicle Manufacturers (RDA) and Porsche GmbH, with the stiff requirement for Porsche to produce a working prototype within ten months!

The first VW saloon was unveiled on 5 February 1936, and contained some sensational features. The chassis, for instance, had independent suspension with torsion bar springing and friction shock absorbers. Air-cooled engines, with two- or four-stroke cycles, developed 22 bhp and were interchangeable; their soft rubber mountings, too, were a considerable advance. And, although there were still no hydraulic brakes in sight, a servo effect was obtained through eccentrics in the cable braking system.

Trials began in October 1936, with three Volkswagens (literally 'people's cars') of the type VW3 equipped with various monitoring devices. By Christmas these three prototypes, driven in two shifts, had each covered 50,000 km (31,068 miles), filling a servicing log book. The findings were incorporated into the next thirty test models to be built, which as a batch had the code-name 'VW 30'. During 1937 this fleet covered well over 1¼ million miles (2 million km) on test.

In 1938 another test batch, 'VW 38', gave the car its final shape. With 24 bhp from the air-cooled Porsche engine, the Volkswagen had a top speed of 65 mph (105 km/h) and was proved fit for continuous top-speed operation on the new German autobahn system.

In parallel with the development of the Volkswagen, work had begun on a new manufacturing plant at Wolfsburg. The topping-out ceremony took place in April 1939, but the outbreak of the Second World War on 1 September stopped both this project and the start of volume car production.

The Volkswagen's production began only in 1945 when 1,785 'Beetles' were produced. Materials were in very short supply, but the pace speeded up considerably in succeeding years. On 4 March 1950 the 100,000th Beetle came off the tracks. The year 1961 was a milestone, seeing the five millionth Beetle produced (and production topped one million in a calendar year for the first time), but on 12 February 1972 the Beetle became a world champion: exactly 15,007,034 examples of Type 1 (as it was called internally) had been produced, finally overtaking Ford's Model T production record. No vehicle before, or since, has been produced in such numbers.

Even though the Beetle has been produced only in America (Mexico and Brazil) since 1977, it is still alive and kicking, and production is now beyond the twenty million mark.

Volkswagen's experimental 'VW 3', on test from October 1936. The 1-litre engine produced 24 bhp, and the car's maximum speed was around 60 mph (100 km/h).

The later 'VW 30' series of Volkswagens was built by Daimler-Benz. The thirty cars were thoroughly tested from 1937, covering more than a million miles. Specifications were the same as for the VW 3.

Wanderer: Type 7

Ferdinand Porsche, together with four engineers, founded the 'Prof Dr Ing hc Ferdinand Porsche' office on 1 December 1930, at Kronenstrasse 42, Stuttgart.

This independent engineering consultancy's first order came from the Wanderer company, requiring a medium class saloon which would rival the Mercedes 170: Porsche's design number for this project was K7.

Wanderer designated the designs W21 and W22, the differences mainly being in the power units. Both had six-cylinder in-line engines, W21 of 1,690 cc (65×85 mm) developing 35 bhp, W22 with 1,950 cc (70×85 mm) and producing 40 bhp.

An experimental Wanderer W10 of 1930/31, with the new six-cylinder Porsche engine.

Wheel-loader: 80/120/200 bhp

Between May 1964 and October 1967 Porsche developed a series of wheel-loaders in three power stages, of 80, 120 and 200 bhp. The scope of this project covered market analysis, conceptual studies, initial drawings, design, procurement, construction of prototypes, and preliminary trials.

The main features of the wheel-loader series included hydraulically controlled articulation, power shift gearbox with four forward and reverse gears, a hydraulic torque converter, one-hand operation for driving and shovel control, and provision for extended operating periods.

Porsche's wheel-loader design went right through to the trials stage, although it never went into production.

Wind-power generators: Types 135/136/137

Although the Porsche name is synonymous with mechanical propulsion, Professor Porsche's interest in wind power, and electrical generators, went right back to the turn of the century. In 1940 Porsche was awarded a contract by the Research Institute for Wind Power Generation (Berlin), to develop wind-power generators in three categories. The smallest, Type 135, was a twin-blade unit mounted on a wooden mast, and produced 130 watts.

A higher-capacity generator, Type 136, produced 736 watts, using a four-blade rotor with adjustable blades regulating power and speed, and protecting the machine against storms. Blade pitch adjustment was carried out by means of a servo-adjuster mechanism actuated by a pressure screw and a centrifugal governor.

The most powerful generator in the family, Type 137, continued in development after the war, and was intended to produce 10,000 watts. The overall diameter of the blades was 30 ft 2 in (9.2 m), each blade being 13 ft 1 in (4 m) in length, 25½ in (650 mm) wide, and weighing 154 lb (70 kg). The blades were made of sheet metal with a stressed outer skin, and were adjustable for pitch.

A pair of spur gears, giving a total transmission ratio of 19.3:1, was used to convey the windmill power to the electricity generator. This dc motor was connected mechanically to a regulating dynamo for matching to the windmill characteristics.

The units were mounted in a lightweight fuselage with directional fins mounted on a 55 ft 9 in (17 m) high tower. A building at the base of the tower housed the switchgear and a 280 Ah buffer battery to complete the installation.

The test rigs were put through their paces and produced good results, one supplying a medium-sized farm in the Stuttgart area with electricity before 1945. Only recently, though, in 1983, was the concept taken any further—the 'Growian' system in Germany has blended Porsche's ideas and research with modern technology, as a start to reducing the country's dependence on oil.

The largest of the generators, Type 137, with a 30 ft 2 in (9.2 m) overall diameter, seen during Porsche tests in 1944.

The Porsche generator Type 136, producing 736 watts, supplied a farm at Hohenheim, near Stuttgart, with electricity from 1944.

World cycling record: 935 base

A Porsche 935 driven by the Frenchman, Henri Pescarolo, was used in 1978 as a pace-maker in an attempt by Jean-Claude Rude to achieve a new world cycling speed record.

Pace-making vehicles have always been used by racing cyclists in record attempts in order to beat air resistance. As far back as 1899 the American, Charles Murphy, had reached 63.25 mph (101.78 km/h) behind an express train! Half a century later the next barrier fell when the Frenchman, José Meiffret, using a special bicycle, reached 127.27 mph (204.778 km/h) on the autobahn near Freiburg behind a Mercedes 300 SL.

It was another Frenchman, Rude, who wanted to take the record above 155 mph (250 km/h), and to do so he and Pescarolo persuaded Porsche's engineers to build a special cage at the back of a factory racing 935.

During initial trial runs on the Volkswagen test track at Ehra-Lessin, 105 mph (170 km/h) was reached straight away. To get there Rude had to overcome the problem of impossibly high gearing, by getting a motor cycle rider with a lance to push him along until he could take over with the pedals! Each revolution of the chain wheel represented a distance of 29½ yd (27 m).

Despite all efforts, a suitable stretch of autoroute could not be found in France, so it was decided to make the record attempt at VW's test ground where the trials had been carried out.

When the run took place in August 1978 Rude had bad news and good news: the cycle's rear tyre burst at around 105 mph (170 km/h) . . . but he was able to avoid a spectacular crash. Rude decided to follow this up with a bid for the one-hour record, at something over 80 mph (130 km/h), but again a tyre failure nearly spelled disaster. On both occasions the glued-on rubber tyre had become detached from the rim, and Rude decided at this point to abandon any further attempts.

Jean-Claude Rude's attempts at cycle records failed at high speeds due to tyre defects.

The attempt on the world cycling record in early 1978 was made behind a Porsche 935.

The 935 was specially equipped with an aerodynamic shield.

World-record cars: Types 910/917-30

In the 1950s and 1960s hill-climbing was nearly as important to Porsche as circuit racing, and the firm was usually successful in the European Mountain Championship. At the end of the 1965 season Porsche's PR and Sport director, Huschke von Hanstein, took a special Type 910 Bergspyder, with a 240 bhp eight-cylinder engine, to the Hockenheimring for an assault on various short-distance world records.

By removing the alternator and rear brakes, and using a motor cycle battery, the car's weight was reduced to 1,070 lb (485 kg), and, despite very wet conditions, von Hanstein set new FIA-recognized world records over a quarter of a mile and 500 metres. The quarter-mile was covered in 11.892 seconds, with an average of 121.97 km/h (75.79 mph), and 500 metres in 13.557 seconds, an average of 132.77 km/h (82.5 mph). Both records were also recognized as international class records in Class E, up to 2 litres capacity.

Since these record attempts had gone so well, von Hanstein then tackled the world record for the standing 1,000 metres, which he covered in 22.212 seconds at an average of 162.074 km/h (100.71 mph)—the terminal speed, incidentally, was 240 km/h (150 mph). As a matter of interest, the world record for the standing kilometre stood to Bernd Rosemeyer in the Auto Union Grand Prix car, which had a 6-litre capacity and 600 bhp.

A good ten years later an entirely different world record was set by Mark Donohue in the Porsche 917-30, illustrating the major technical advances made in this period.

For the Canadian-American Challenge (Can-Am) series Porsche had developed the Type 917 racing car, which reached the peak of its development in 1973 as the 917-30 model. The superiority of the car, with its 5.4-litre twelve-cylinder turbocharged engine, was so great that the organizers were only too happy to change the rules for the following year, taking heed of the energy crisis. With exhaust turbocharging the engine produced 1,100 bhp, making it the most powerful engine ever constructed for circuit racing, and it was installed in a chassis weighing a mere 1,760 lb (800 kg).

Although the change of rules for 1974 effectively ruled out the 917-30, regular driver Mark Donohue, winner of the 1973 Can-Am series, wanted to establish it as an all-time great before it disappeared into a museum. He had his eyes on the closed-circuit world-record speed, held by A.J. Foyt in a USAC formula Coyote, at 350.53 km/h (217.82 mph).

Donohue chose the oval track at Talladega, in Alabama, for his attempt, but two engines failed while testing for the record. He had virtually given up when Porsche sent over a special 5-litre engine, with a new intercooler to lower the temperature of the charged air. This proved perfectly successful and Donohue set his world record on 8 August 1975, at 355.85 km/h (221.12 mph). The elation was sadly short-lived, for ten days later Mark Donohue lost his life in the Formula 1 race at the Österreichring.

Huschke von Hanstein set new world records with the 910 Bergspyder in 1965. The 2-litre, eight-cylinder car developing 240 bhp covered 500 m from a standing start in 13.5 seconds, averaging 132.7 km/h (82.5 mph).

Mark Donohue pictured at Talladega, Alabama, on his way to a new closed-circuit world record of 355.85 km/h (221.12 mph) in 1975. The car developed 1,100 bhp, and special wheel discs were fitted to reduce drag.

Zimmer: 910 base

When the Sports Car Club of America introduced a 2-litre limit for sports car racing, early in the 1970s, Jack Zimmer had the choice of fitting a smaller engine into his (2.2-litre) Porsche 910, or withdrawing it from competition. He decided to convert the racing car into a stylish, exciting machine suitable for use on the public highways.

Zimmer contacted Charles Pelly, a freelance designer, who built several scale models, and Dick Troutman in California was then given the job of carrying out the body conversion.

Aluminium was chosen for the body skin, and the angles and curvatures were clearly going to call for a high degree of skill. The conversion took 14 months to complete, and cost the equivalent of DM 100,000. Ironically, the four-cam eight-cylinder engine was taken out, and replaced by a 2-litre 911 engine with fuel injection, tuned to over 200 bhp; the car's top speed was estimated at 136 mph (220 km/h).

The conventional tubular frame of the 910, built for racing in the 1967/68 seasons, was retained, along with the original Porsche racing five-speed transmission, and limited slip differential, located behind the engine. Twin aluminium fuel tanks with a total capacity of 19.8 gal (90 litres) were used, and the standard ventilated disc brakes also featured. The wheel rims were 9.5×13 at the front and 13×13 at the rear, equipped with Firestone tyres ($500/950 \times 13$ front, $600/1300 \times 13$ rear).

The aluminium body consisted of three large sections—the front, the cockpit and the rear—though the roof was made of steel. The windows were taken from the Cadillac range and altered, and standard Porsche 911 instruments were used.

Jack Zimmer and his racing Porsche 910 during its conversion to road car trim. The futuristic car was reskinned in aluminium, and the job took fourteen months to complete.

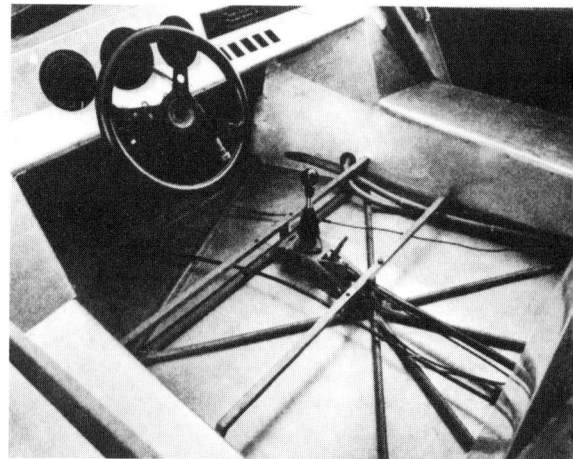

The 910's tubular frame was retained, although the interior fittings were far removed from the normal racing equipment.

The side and rear views of the Zimmer 910 were quite unusual, and gave no hint of Porsche ancestry. The eight-cylinder engine was replaced by a tuned 911, 2-litre, six-cylinder unit developing over 200 bhp.

Weissach: the ideas factory

Initially, the site acquired by Porsche near the village of Weissach was intended for the construction of test tracks, starting with a circular steering pad 220 yd (200 m) in diameter. The first soil was turned by Dr Ferry Porsche (using a bulldozer) on 16 October 1961.

A young team of engineers carried out the planning and construction of the test area, which by 1970 included a racing circuit, together with linking road systems needed for operational and durability trials. Since 1954 steering pad tests had been carried out at Malmsheim, on a former military airfield, but, as Weissach developed, more and more of Porsche's testing could be carried out there.

When the shortage of space at Zuffenhausen became even more acute, as a result of increasing orders, it was decided to transfer the entire development department, including the test beds, to Weissach some 15 miles (25 km) away to the west.

Eventually some land became available, initially 103 acres (45 hectares), in the triangle of roads between Weissach, Flacht and Mönsheim, offering ideal conditions for the design and planning departments of Porsche, who used a Swiss firm of consulting engineers to draw up the plans.

Planning of the premises took into account the need for these to be close to the prototype test facilities, and the diversity of vehicles to be tested clearly made considerable demands on the ingenuity of the plant designers, particularly in the areas of electricity supply, ventilation and water supply.

Another consideration was the need for good sound insulation, especially for the racing engines which would be tested over long periods, in view of the proximity of the Weissach villagers a mile away. Large machines would cause vibration, so the buildings would need to have substantial foundations. Obviously, building the test houses involved far more than simply erecting rows of brick boxes!

Prototype manufacture makes use of advanced woodworking machinery, and further models are then shaped in clay, and by hand or machine in a light metal foundry. The capacity for this three-man operation is about 660 lb (300 kg) of materials daily. Two forging hammers are needed for the manufacture of test forgings, and a hardening shop supplements these auxiliary operations.

Engine manufacture takes place in the mechanical workshop, with facilities for parts inspection. In addition to the conventional machine tools, there are two computer-controlled boring mills on site, along with lathes and grinding machines for crankshafts and camshafts, and a sophisticated measuring machine.

Two workshops are available for prototype body manufacture: the body shop contains sheet-metal working machines and equipment to work aluminium and magnesium; the lightweight bodies for racing cars, on the other hand, are produced in the plastics body shop using the latest in glass-fibre-reinforced materials. Occasionally this workshop carries out special runs

The test houses at Weissach, built in the early 1970s.

for the production line at Zuffenhausen, the 500 'ducktail' rear spoilers for the Carrera RS being an example.

A combined sanding, spraying and drying area in the same building allows vehicles to be sprayed with stoving enamels at temperatures of up to 80°C. A small electrical and upholstery department completes the bodywork section.

In the northern building, engines, gearboxes, chassis and bodies are assembled. Two test bays check wheel alignment, while engines and brakes are dynamically tested in the vehicle on a roller dynamometer, prior to the road test.

The first bench-test building contains ten engine test-beds for water, or air-cooled power units. The test bays are designed for running air-cooled engines of up to 500 bhp, and, in order to maintain constant atmospheric conditions, the engine intakes are supplied, from the air conditioning system, with air at 20°C (68°F) and 60 per cent relative humidity.

All test data are transferred for analysis to a centralized computer, which also controls progressive bench-test performance measurements. Four test cells with roller test stands are used for exhaust analyses on complete cars, and the mounting of these rollers on springs proved successful in reducing vibration within the building. As early as the end of 1970 Porsche had carried out continuous running exhaust gas tests over 50,000-mile cycles, all on programme computer-controlled rollers.

Additionally, four roller tests were set up outdoors to maintain the normal mix of weather conditions. Cars are accelerated, braked, and change gear completely automatically, as a paper tape programme controls the emission cycle operations, eliminating the human factor . . . and relieving the staff of an extremely boring job, to boot. The only person in sight is the supervisor, who fills the cars up with fuel from time to time! In these cars, the pedals are operated by a servo-motor, and all other operations including gear-changing are controlled by a pneumatic cylinder. The mechanisms—all Porsche's own developments—fit in place of the driver's seat and can be adjusted on the seat runners.

Anechoic room for noise measurements . . . it's uncannily quiet in there!

Porsche's in-house foundry produces parts for prototypes, and small series production.

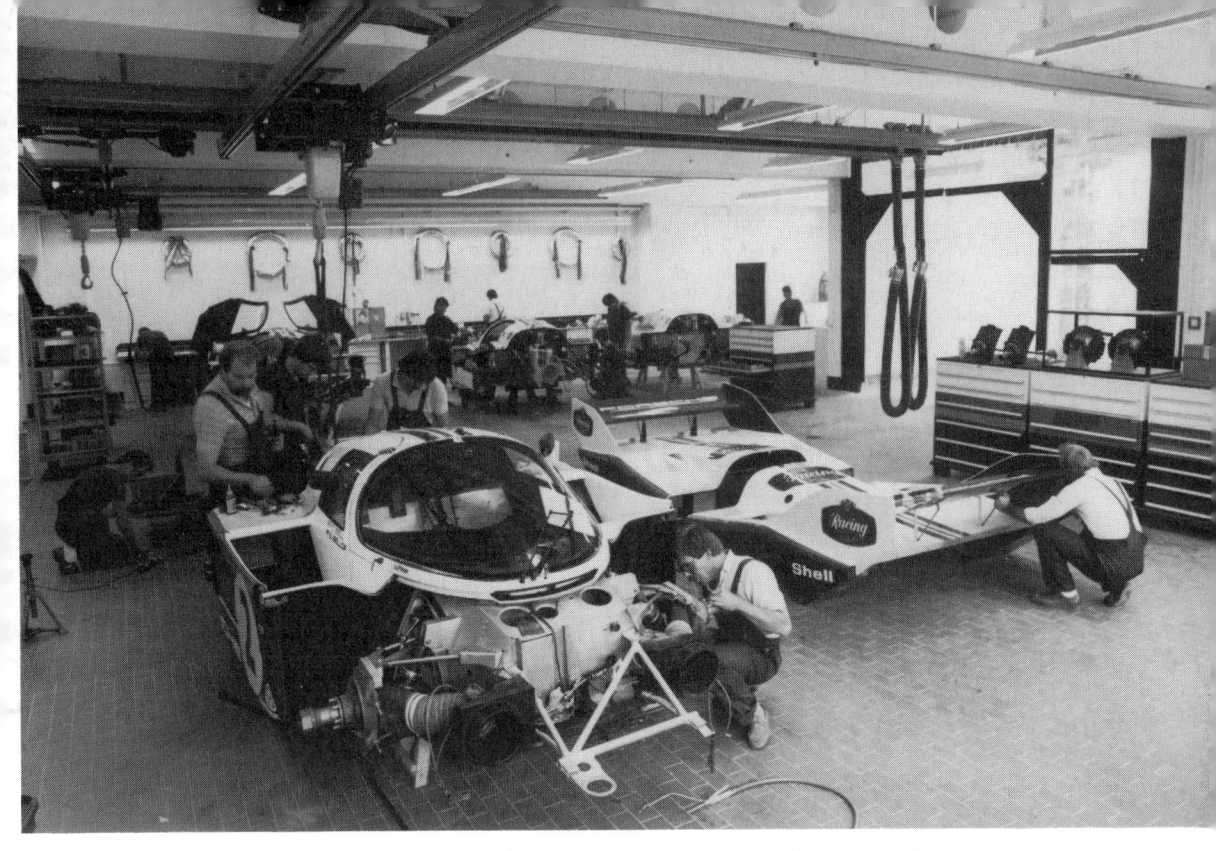

Body assembly in the race department. In the foreground is the makings of Bell's and Bellof's 956.

The engine shop, showing assembly of a TAG Turbo Grand Prix engine in progress.

A wind tunnel, built of insulated steel, is used for experiments with heating, ventilation and air conditioning in vehicles. Wind velocities of up to 105 mph (170 km/h) can be generated, with temperature control of between −40°C and 70°C (−40°F and +158°F).

A drum-type test bed, 6½ ft (2 m) in diameter, facilitates the testing of front-, rear-, or four-wheel-drive designs. The drive can be switched from front to rear, or to both axles at the same time, at speeds of up to 250 mph (400 km/h) and with outputs of up to 600 bhp. This test bed has also been used for racing car testing, including the Can-Am 917 turbo which initially achieved 850 bhp and could accelerate at more than 1g.

In order to combat vehicle and engine noise an anechoic test chamber was one of the original installations, equipped with a drum-type test bed. This chamber is supported in isolation from the building by springs, and insulated from the environment with mats, while internal insulation is provided by glass-fibre cubes.

A crash 'sledge' is used for safety tests on vehicles or mock-ups, propelled by a high-pressure cylinder with a thrust of up to 120 Mp and with an acceleration of up to 60 g. Every stage of the tests is recorded on high-speed cameras.

As well as these test beds, there are smaller units for experiments with braking systems, chassis and structural vibrations, transmission systems, and tyres. Then the materials-testing laboratory runs static and dynamic component and fatigue test cycles using the appropriate specialized machinery. A photographic laboratory with the latest equipment, and an electron-beam welding machine, complete this department.

The natural terrain provided ideal conditions for the creation of various proving ground tracks. These conditions were fully utilized by the building of link roads to produce a single course, combining high-speed sections with a hill-climb, for instance. For suspension testing, tracks include potholes, washboard, Belgian pavé, a rail crossing, a salt water section, transverse grooves, jumping board and steep sections.

Above *Stand clear! The 'sledge' used for the crash testing programme in one of the main halls.*

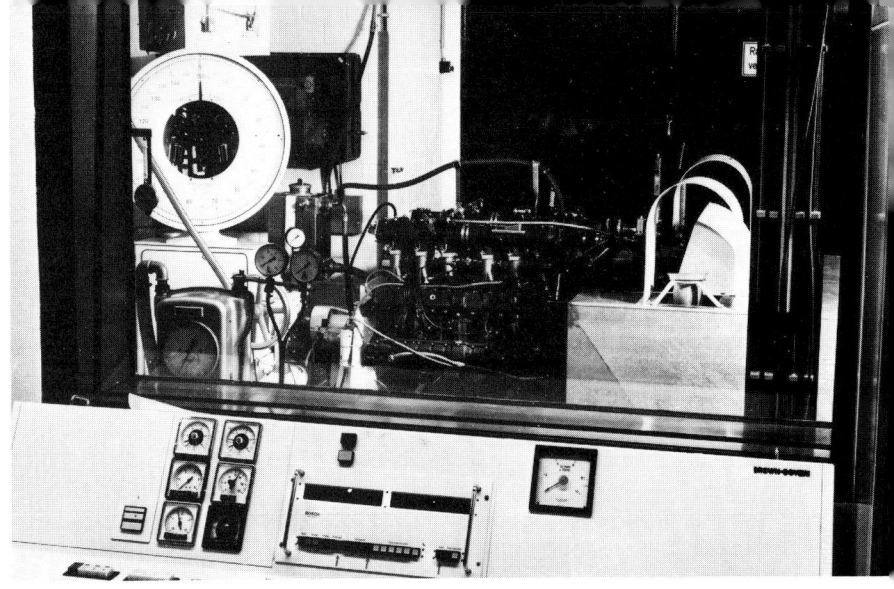

Top right *Dynamometers in a sound-proofed glass cabin can cope with outputs of from 55 to 1,000 horsepower.*

Above right *Transmission test beds for standard gearboxes and transaxle systems.*

Right *Test stands for experiments with different braking systems.*

Even the racing cars have to run on this destructive track as part of their development!

Within the area is a full cross-country course—ideal for tank-testing but also used for rally car development—and the skid pad with diameters of 40, 60 and 190 m (130, 195 and 625 ft). The largest, outer circuit is also used for aquaplaning tests on a section which can be flooded to order, and measuring equipment is integrated.

The first buildings were occupied early in the 1970s and had, already, cost DM 80 million. Nearly 1,000 people worked there originally, many recruited from Zuffenhausen but more from the surrounding district. Ten years later 1,800 people were employed at Weissach, which was literally bursting at the seams, and a great deal of development is going on all the time. A fully equipped exhaust emission laboratory is now on site, and the competitions department has been rehoused in the centre of the complex in new, modern buildings: here the company's racing cars are designed and built, alongside the TAG-Porsche Formula 1 engines. By mid-1985 the total number of people employed at Weissach had risen dramatically to 2,400 people, nearly one-third of Porsche's total payroll, reflecting not only the considerable amount of in-house research and development, but also the value of outside contracts. Of this Weissach workforce, around 100 are employed in the competitions department. Even the hexagonal administration block is changing shape as new sections are added, many offices having been moved to yet another new building. Car parking, too, has to be constantly expanded.

Professor Porsche's design studio, founded in 1930 with four principal employees, has now developed into an ideas factory, and its output is in heavy demand from all parts of the world. The previous section in this book, detailing a few of the Porsche-inspired developments, is only the tip of the iceberg so far as current and future research are concerned.

An aerial view of Weissach, showing the road and off-road courses, the original steering pad, and the test and headquarters buildings. Even more construction work has been done since this photograph was taken.

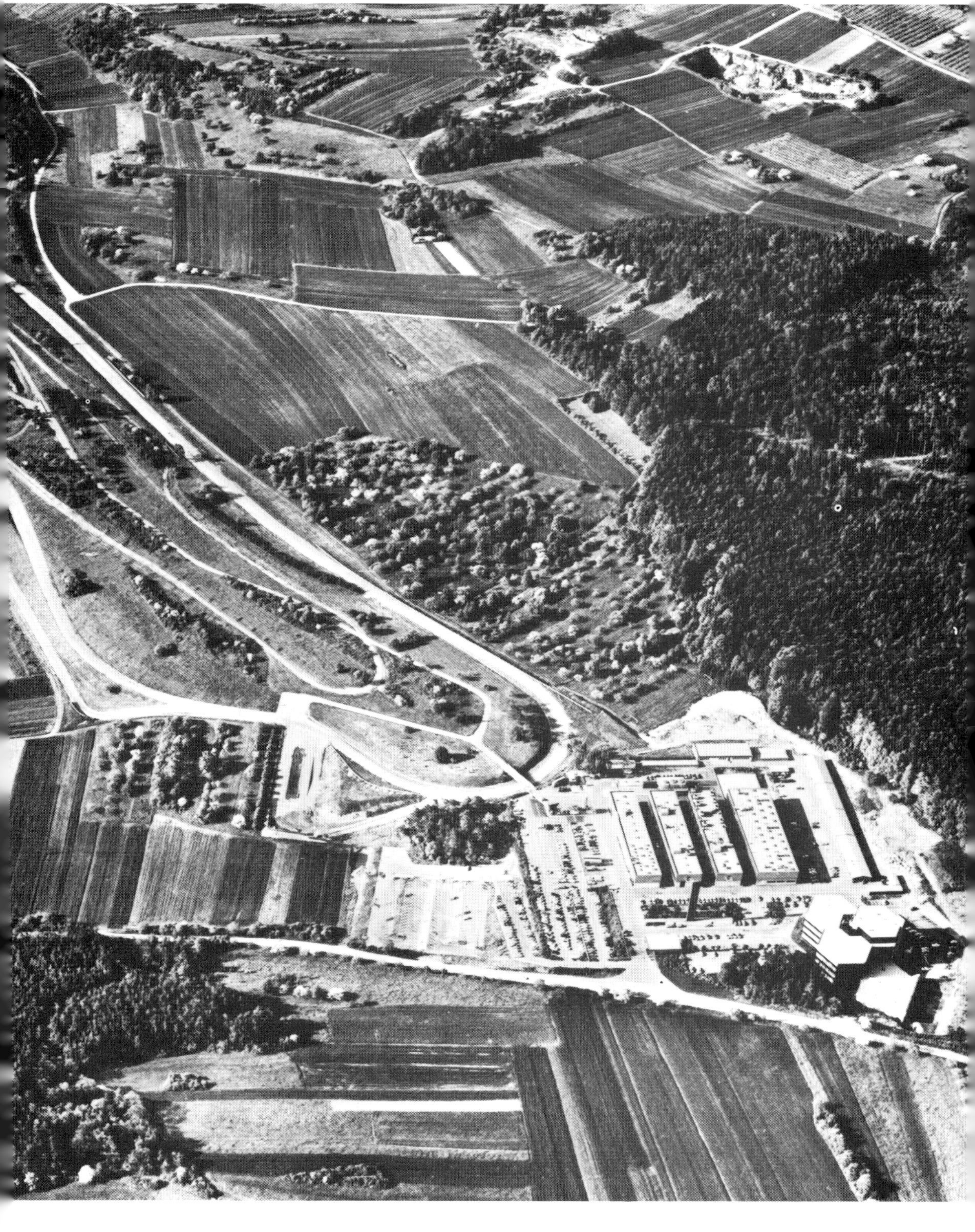

APPENDICES

1: Various special types and designs

Baker	914-6 Turbo
Bauer	356-base special
Brendel	356-base special
Buggy	356 engineering
Corvair	550-base special
Dragster	Porsche engines
Koch	911-base special
Morse	924 Cabriolet
Racing classic	718 replica
Ruthan	550-base special
Stephani	356-base special
Tour de France	904 base

Of the 'specials' listed here, there were unfortunately only incomplete documents and data available. If any readers have further information, including addresses, they are invited to contact Lothar Boschen via the publisher.

Baker: A Porsche 914-6 equipped with a turbocharged 3-litre engine for American NASCAR racing.

Bauer: *A customized Porsche 356.*

Brendel: *A 356A with a Brendel hardtop seen at the 1959 Gaisberg race meeting.*

Buggy: Built by Porsche-Salzburg, this buggy uses Porsche's engine and chassis.

Corvair: A Porsche 550 Spyder, equipped with an American Corvair engine.

Dragster: One of the many Porsche-powered American dragsters.

Koch: A 1970s' Targa with three front seats.

Morse: A complete Cabriolet based on the 924 (or 944) model, from an American workshop.

Racing Classic: A replica of the famous 718 Spyder, based on a Volkswagen.

Rhutan: *An individually designed modification of a Porsche Spyder 'special' in America . . . wins no prizes for styling!*

Stephani: *Special paintwork and enclosed rear wheels from a Swiss company, based on the 356.*

Tour de France: An uprated Porsche 904 produced for the 1971 Tour de France event.

2: Porsche type number index

7 Wanderer: 1.7-/1.8-litre six-cylinder saloon.
8 Wanderer: eight-cylinder coupé (3,250 cc engine).
9 Wanderer: Type 7 with streamlined body; prototype.
10 Horch: independent rear suspension.
12 Zündapp: five-cylinder radial engine for rear mounting in saloon; three prototypes (forerunner to VW Beetle).
16 Roehr: eight-cylinder engine.
18 Air-cooled lorry engine, 3,500 cc radial unit.
19 Triaxial variation on Type 18.
20 Teves: torsion bar springing/suspension.
22 Auto Union: Grand Prix car—sixteen cylinders, mid-engine, 4,360 cc.
24 Zündapp: three-wheel vehicle.
27 Mathis-Ford: saloon.
28 Type 18/19: modification.
30 Hanomag: independent front suspension.
31 Wanderer: independent front suspension.
32 NSU: four-cylinder flat-type engine, air-cooled for rear mounting in saloon; three prototypes (VW Beetle forerunner).
36 Roehr: supercharged eight-cylinder engine and chassis.
38 Modification of a petrol engine for two-axle lorry.
39 Modification of a petrol engine for three-axle lorry.
52 Sports car: drawing based on Type 22.
56 ERA: front suspension.
57 Zündapp: motor cycle engine.
59 Rochet-Schneider: independent front suspension.
60 Volkswagen/KdF car: 22 bhp, air-cooled rear engine, rear-wheel drive.
61 Volkswagen: narrower version of the Type 60.
62 Volkswagen: Kübelwagen (bucket car) prototype.
64 Sports car: streamlined shape—based on VW60 (also type 60 K10), three prototypes.
65 Volkswagen: driving school fittings.
66 Volkswagen: right-hand drive.
67 Volkswagen: invalid carriage.
68 Volkswagen: 'Reichspost' (Post Office) delivery van.
70 Aircraft engine: 32 cylinders.
72 Aircraft engine: sixteen cylinders.
76 Laboratory unit.
78 Aircraft engine: slide valve gear.
80 Daimler-Benz: world-record car, three axles, 2,200 bhp, up to 370 mph (600 km/h).
81 Volkswagen: medium van chassis (body 286).
82 Volkswagen: Kübelwagen (bucket car), 1,130 cc, single-axle drive, self-locking differential.
83 Volkswagen: 'Kreis' power transmission system.
84 Volkswagen: 'Dr Hering' power transmission system.
85 Volkswagen: four-wheel drive study.
86 Volkswagen: Kübelwagen (bucket car) with four-wheel drive (study).
87 Volkswagen: four-wheel drive for Kommandeur car (Type 82).

88 Volkswagen: modified bus on Kübelwagen chassis.
89 Volkswagen: 'Beier' automatic gearbox system (study).
92 Volkswagen: cross-country version of Type 82.
93 Limited-slip differential.
94 Daimler-Benz: racing engine.
95 Bus suspension/chassis.
96 Hydraulic power transmission.
97 Daimler-Benz: heavy lorry.
98 Volkswagen: amphibious car (Type 128) with Type 62-CL body.
99 Study for Types 91 and 95.
100 Leopard tank with 7.5 mm gun, two air-cooled V10 engines (10-litre capacity) and electric gear switching.
101 Tiger tank with 15-litre twin engine and electric transmission using mixed mode.
102 Tiger tank with Voith electric transmission.
103 Tiger tank with Voith hydraulic transmission.
106 Volkswagen: PIM experimental power transmission (Type 60).
107 Volkswagen: supercharged engine.
108 Daimler-Benz: two-stage supercharged engine.
109 Two motor cycle engines.
110 'Volkspflug' (people's plough) small tractor.
111 Tractor with two-cylinder engine.
112 Improved version of Type 111.
113 Tractor with 15 bhp two-cylinder in-line engine.
114 Sports car: study with 1.5-litre (72 bhp), V10 twin camshaft engine.
115 Volkswagen: VW engine with supercharger, overhead camshafts and hemispherical combustion chambers (base: 1.1-litre).
116 Volkswagen: study for a racing car based on Type 114.
120 Volkswagen: engine as standby power supply (RLM).
121 Volkswagen: stationary VW engine (HWA).
122 Volkswagen: stationary VW engine with battery ignition (RP).
124 Kübelwagen: modified for use on rails.
125 Study for a wind generator.
126 Volkswagen: fully synchronized power transmission.
128 Volkswagen: amphibious car (Type 87), 1st version.
129 Volkswagen: short chassis version of the Type 128.
133 Carburettor, naturally aspirated.
135 Wind-power generator, 130 W.
136 Wind-power generator, 736 W.
137 Wind-power generator, 4,550 W.
138 Volkswagen: amphibious car (Type 87), 2nd version.
139 Volkswagen: amphibious car, 2nd version (Type 138 S), without centre frame.
151 Volkswagen: 'Plus' power transmission system.
152 Volkswagen: 'Stieber' power transmission system.
153 Skoda: Ostrad traction engine, air-cooled six-cylinder engine.
155 Type 82: half-track vehicle version.
156 Type 166: version for rail truck.
157 Type 82/Type 87: version for rail truck.
158 Wood-gas generator.
160 Volkswagen: self-supporting body (investigation).
162 Volkswagen: Kübelwagen with self-supporting body.
164 Cross-country lorry with six wheels and two engines.
166 Volkswagen: amphibious car (production model).
170 Landing-craft engine: based on VW engine with 40 bhp (model 1).

171 Landing-craft engine (model 2).
174 Landing-craft engine (normal version).
175 Ostrad traction engine: military traction engine with steel wheels.
177 Type 82/Type 87: five-speed gearbox.
179 Volkswagen: injection system for VW engine.
180 Tiger tank with petrol engine and electric transmission.
181 Type 100: hydraulic transmission.
182 Volkswagen: Kübelwagen (production model).
187 Type 182: four-wheel drive.
198 Tank starter gear for Type 82.
200 Diesel engine: 10 litres capacity, air-cooled.
205 'Maus' tank with 15 cm gun, 1,080 bhp, 44.5-litre diesel, electric transmission.
209 Daimler-Benz: 44.5-litre diesel engine, V12, air-cooled.
212 Sixteen-cylinder diesel, air-cooled, basis for Type 205.
220 V16 diesel, air-cooled, supercharged, 36.8-litre.
225 Volkswagen: electric power transmission from Brown, Boverie & Cie.
230 Volkswagen: wood-gas generator.
231 Volkswagen: acetylene mixture.
235 Volkswagen: electric transmission.
236 Volkswagen: grating for use with Type 230.
238 Volkswagen: cable hoist.
239 Type 82: wood-gas generator.
240 Volkswagen: propane gas operation of VW engine.
245 18-ton multi-purpose tank (drawings, models).
247 Aircraft engine: based on VW engine.
250 Turretless tank with 10.5 cm gun.
252 Volkswagen: 'PIV' power transmission system.
261 Heating for Panther tank.
276 Type 82: pintle hook.
283 Type 82: design for wood-gas generator.
285 Water turbine, 3.5 bhp.
287 Kommandeur car on four-wheel drive chassis.
288 Water turbine, 13 bhp.
289 Water turbine, 15 bhp.
291 Wind tunnel, 600 mm (23.6 in).
292 Wind tunnel, 300 mm (11.8 in).
293 Caterpillar tractor (for wheel or rail use).
294 Ski binding.
298 Radio receiver for Volkswagen.
307 Dense-medium carburettor.
309 Two-stroke diesel engine for Volkswagen and/or tractor.
312 Tractor with petrol engine.
313 Type 312 with 17 bhp air-cooled diesel engine.
315 Ski cable for VW engine.
323 Tractor with 11 bhp diesel engine.
328 Tractor, 28 bhp.
335 Winch.
336 Capstan winch.
339 Conveyor belt.
340 Two-wheel wheelbarrow.
352 von Senger: saloon (study).
355 Volkswagen: delivery van based on Type 81/83.
356 Porsche sports car: mid-engine roadster, series 356, A, B, C built between 1947 and 1964.
360 Cisitalia: Flat twelve-cylinder engine, air-cooled, supercharged, 1.5-litre, switchable four-wheel drive.
361 Type 360: single-cylinder test engine.
362 Type 360: 2-litre atmospheric engine.
369 Porsche: 1.1-litre engine.
370 Cisitalia: 1.5-litre sports car (study).

372 Cisitalia: 2-litre sports car, air-cooled, 100 bhp, V8 engine, five-speed gearbox, five-seater (study).
383 Porsche: synchromesh system for VW gearbox ('Schmid').
384 Porsche: alternative to 383.
385 Cisitalia: water turbine.
425 20 bhp diesel tractor.
502 Porsche: 1.5-litre engine, Hirth crankshaft, 55 bhp.
506 Porsche: 1.3-litre engine (506/0 = 1,286 cc, and 506/1 = 1,290 cc, as well as 506/2 = three-part crankcase).
509 Porsche: 1.3-litre prototype (from 506 series?).
514 Porsche: works racing car for 1951 Le Mans 24 h.
519 Porsche: synchromesh four-speed gearbox (from VW gearbox).
522 Porsche: strut-type front suspension combining VW trailing links with MacPherson strut. Test car known as 'Storch'.
524 Porsche: 1.3-litre with Bosch fuel injection.
527 Porsche: 1.5-litre for Le Mans (later series).
528 Porsche: 1.5-litre, as 528/1 with two-part crankcase and 528/2 with three-part crankcase.
530 Porsche: four-seater model with 2,400 mm (94½ in) wheelbase.
531 Porsche: 1.3-litre with new camshaft.
532 Porsche: single carburettor for Type 369.
533 Porsche: 1.1-litre racing engine.
534 Volkswagen: small sports car.
539 Porsche: 1.5-litre.
540 Porsche: American Roadster.
541 Porsche: special sports version of Type 356.
542 Studebaker: air- and water-cooled V6 engine.
543 Porsche: 1.5-litre industrial engine.
544 Porsche: 1.5-litre industrial engine.
546 Porsche: plain-bearing version of Type 527 (1500N), as 546/2 with three-part crankcase.
547 Four-camshaft 1.5-litre engine (547/1) in production version. Type 718 and 718/2 (542/2) each 1.5-litre; 1.6-litre racing engine for Type 718 (547/4); 1.7-litre racing engine for 718 (547/5).
550 Porsche: Spyder, two-seater racing sports car. 550A with tubular lattice frame.
555 Volkswagen: prototypes.
557 Porsche: 1.5-litre for USA.
559 Porsche: power transmission (study).
568 Porsche: exhaust-induced cooling ('Fletcher').
574 Porsche: electrical clutch for Type 356.
575 Porsche: support bracket (production).
577 Porsche: disc brakes for Type 356 (Dunlop).
587 Porsche: 2-litre version of Type 547.
588 Porsche: power transmission for Type 587.
589 Porsche: 1.3-litre S-engine for Type 356 with two-part crankcase 589/2: three-part crankcase.
592 Porsche: 2-litre (?).
593 Porsche: four-speed gearbox (?).
596 Porsche: two-cylinder industrial engine.
597 Porsche: Jagdwagen ('hunting car') with four-wheel drive (1.5-litre).
606 Volkswagen: 1.5-litre underfloor engine.
616 Porsche: 1.6-litre for Type 356A; 616/2 with 1.6-litre (1600S); 616/3 with 1.6-litre (industrial engine); 616/7 with 90 bhp (356 B); 616/12 for 616/2 with cast-iron cylinders for 356 B; 616/15 for 1.6-litre of Type 356 C (1600 C); 616/16 for 1.6-litre (1600 SC); 616/36 for 1.6-litre Porsche 912; 616/39 for 912 engine with US

exhaust detoxification.
619 Volkswagen: diesel engine studies.
621 Porsche: single-cylinder tractor (Allgaier).
622 Porsche: two-cylinder tractor.
623 Porsche: three-cylinder tractor.
624 Porsche: four-cylinder tractor.
627 Volkswagen: swing axle with strut position on the frame.
628 Volkswagen: fresh air heating.
631 Porsche: diesel engine (studies).
632 Porsche: development studies for Type 356.
633 Studebaker: saloon.
638 Volkswagen: study for V6 engine, 1.2- and 1.6-litre.
643 Porsche: four-speed gearbox for Type 356.
644 Porsche: modifications from 356 to 356 A chassis, with barrel-type gearbox housing.
645 Porsche: Spyder, prototype 'Mickey Mouse' under R. von Frankenberg.
654 Motor boat study.
655 Porsche: 50 cc moped engine.
672 Volkswagen: small car with underfloor engine (study).
673 Volkswagen: six-cylinder underfloor engine, 1.2- and 1.6-litre.
675 Volkswagen: small car.
678 Aero-engines: 678/1 for 65 bhp with step-down gearing, 678/3 for 53 bhp with direct drive; 678/4 for 75 bhp with step-down gearing.
690 Porsche: five-speed gearbox for RSK 718 (racing version of the 644 barrel-type gearbox).
691 Porsche: Spyder for 1956 (550 RS).
692 Porsche: four-camshaft engine in modified version. 692/0, 1.5-litre with roller-bearing crankshaft; 692/1 for plain-bearing crankshaft; 692/2 for Carrera series; 692/3 for sports version of the 692/2 and as 692/3A with camshaft balancing weights.
693 Porsche: 1.3-litre based on Type 547.
694 Porsche: cross-country version as study for Type 597.
695 Porsche: predevelopment (T7) for Porsche 911.
700 Volkswagen: large-capacity car (study).
702 One-man helicopter with 678/3 engine.
703 Porsche: 1.6-litre experiments.
704 Porsche: diesel engine.
709 Porsche: gearbox studies.
710 Porsche: gearbox for Type 356 with improved synchromesh.
715 Porsche: test-bed gearbox.
716 Porsche: annular synchromesh.
718 Porsche: Spyder RSK with mid-engine, 718/2 for 1.5-litre racing car.
719 Porsche: racing engine with fuel injection, successor to the RSK carburettor engine.
722 Porsche: flat-type engine for underfloor mounting (VW Type 3?).
724 Volkswagen: 1.4-litre, flat-type engine.
726 Volkswagen: body on VW chassis.
728 Volkswagen: further development of Type 675.
729 Marine engine, derived from Type 616.
737 Porsche outboard motor.
741 Porsche: four-speed gearbox for 356 B; 741/A for 356 B and C.
742 Porsche: four-wheel drive studies.
745 Porsche: six-cylinder experimental engine, the 2-litre prototype is in the museum.
751 Volkswagen: gearbox with automatic clutch.
752 Volkswagen: flat-type engine, 1,000 cc.
753 Porsche: 1.5-litre, eight-cylinder, for Formula 1.

756 Body and chassis for the special series (20-off) Carrera Abarth.
759 Porsche: four-speed gearbox (inverted drive).
763 Design for series production seat.
764 Volkswagen: six-seater saloon.
768 Porsche: 1.6-litre with fuel injection.
769 Porsche: differential.
771 Porsche: 2- and 2.2-litre, variants of Type 753.
775 Porsche: six-speed gearbox for diesel.
787 Porsche: Formula 1 chassis.
792 Inboard boat engine.
798 Porsche: chassis and transmission for 'Le Mans GT' (2000 GS/GT).
801 Porsche: 1.6- to 1.8-litre opposed-cylinder engine (four cylinders).
802 Four-cylinder fuel injection engine (Michael May?).
804 Porsche: Formula 1, final version for Type 753 engine.
806 Leopard tank.
814 Leopard tank.
820 Volkswagen: VW gearbox with Porsche synchromesh.
821 Porsche: engine for Type 901 (911).
822 Porsche: gearbox for the 771 engine (5/6 gears).
901 Porsche sports car: sold as '911' (T8).
901/00 five-speed gearbox for 911.
901/01 130 bhp engine.
901/02 160 bhp engine.
901/03 110 bhp engine.
901/05 901/01 with Weber carburettor.
901/06 901/05 with improved valve adjustment mechanism.
901/07 901/06 prepared for Sportomatic.
901/08 901/02 prepared for Sportomatic.
901/09 fuel-injection engine for 911 E.
901/10 fuel-injection engine for 911 S.
901/13 901/03 prepared for Sportomatic.
901/14 901/06 (130 bhp) for America.
901/17 901/14 prepared for Sportomatic.
901/20 210 bhp engine for Type 906.
901/21 901/20 as fuel injection unit for Type 906 and Type 910/6.
901/22 210 bhp engine for 911 R.
901/30 150 bhp engine for 911 L rally.
902 Porsche: four-speed gearbox for 911/912.
902/01 four-speed gearbox for 912.
902/02 five-speed gearbox for 912.
902/1 five-speed gearbox for 912/911.
903 Automatic clutch for sports car.
904 Porsche: GT mid-engined coupé with 592 engine.
904/6 six cylinders.
904/8 eight cylinders.
905 Porsche: four-speed Sportomatic gearbox.
905/01 four-speed Sportomatic gearbox (new).
905/13 four-speed Sportomatic, ditto 905/20, 905/21.
906 Porsche: competition car developed from the Type 904.
906/8 eight cylinders.
906/E fuel-injection engine.
907 Porsche: racing coupé with right-hand drive developed from the Type 910, with 901 and 771 engine.
907/L 907 long tail.
908 Porsche: Spyder developed from the Type 907 and eight-cylinder engine.
908/01 coupé with long tail.
908/02 open spyder.
908/03 spyder with altered seating position, developed from the Type 909.
908/K short-tail coupé.

908/L long-tail coupé.
909 Porsche: hill-climb racing car with 2-litre engine.
910 Porsche: racing sports coupé, developed from the Type 906.
910/6 with 901 2-litre engine.
910/8 with 2.2-litre eight-cylinder engine.
910/8 B light version of 910/B (hill-climb car).
911 Commercial designation of the Type 901.
911/00 four-speed gearbox for 911 T.
911/01 2.2-litre engine for 911 E.
911/02 180 bhp engine for 911 S.
911/03 130 bhp engine for 911 T.
911/04 911/01 engine prepared for Sportomatic.
911/06 911/03 engine prepared for Sportomatic.
911/07 2.2-litre for 911 T-USA.
911/08 911/07 prepared for Sportomatic.
911/20 racing engine with 2,247 cc for 911 S.
911/21 racing engine with 2,381 cc for 911 S.
911/22 911/20 in carburettor version.
911/41 2.7-litre for 911.
911/42 2.7-litre for 911 S.
911/43 2.7-litre for 911-USA.
911/44 2.7-litre for 911-California.
911/46 911/41 prepared for Sportomatic.
911/47 911/42 prepared for Sportomatic.
911/48 911/43 prepared for Sportomatic.
911/49 911/44 prepared for Sportomatic.
911/51 2.4-litre for 911-USA.
911/52 2.4-litre for 911-E.
911/53 2.4-litre for 911-S.
911/57 2.4-litre for 911-T.
911/61 911/51 prepared for Sportomatic.
911/62 911/52 ditto.
911/63 911/53 ditto.
911/67 911/57 ditto.
911/70 racing engine, 2,494 cc for 911 S.
911/72 racing engine, 2.8-litre for 911 S/Carrera RSR.
911/73 racing engine, 2,466 cc for 911 S.
911/74 racing engine, 3.0-litre for Carrera RSR.
911/75 911/74 with modified carburettor.
911/76 turbo engine, 2.1-litre for Carrera RSR.
911/77 3.0-litre engine for Carrera RS.
911/81 2.7-litre for 911.
911/82 2.7-litre for 911 S-USA.
911/83 2.7-litre Carrera RS.
911/84 2.7-litre for 911 S-California.
911/86 911/81 prepared for Sportomatic.
911/89 2.7-litre USA engine with Sportomatic.
911/91 2.4-litre with K-Jetronic (911 T-USA).
911/92 2.7-litre for 911 with K-Jetronic.
911/93 2.7-litre for 911 S.
911/96 911/91 Sportomatic.
911/97 911/92 Sportomatic.
911/98 911/93 Sportomatic.
911E fuel-injection model.
911L version for USA.
911R lightweight design for competition.
911S high-powered version of the basic 911.
911T economy version of the 911.
912 Porsche: four-cylinder version of the 911, with engine 616, but also type number for the 4.5-litre twelve-cylinder racing engine of the Type 917.
912 E fuel-injection version of the 912 for USA.
914 Mid-engined sports car (two-seater).
914/6 2.0-litre, Type 901.
914/8 3.0-litre, Type 908.
914/11 five-speed gearbox for 914/6.

914/12 five-speed gearbox for 914.
915 Porsche: higher duty four-/five-speed gearbox for 911.
915/06 five-speed gearbox for 911/911 S and Carrera.
916 Porsche: racing version of Type 901 engine (2 OHC), but also type number for five-speed gearbox for 908/01 and 908/02.
917 Porsche: racing coupé with 4.5-litre, twelve-cylinder engine (Type 912).
917/10 Spyder.
917/20 shortened version with Sera body ('Sau').
917/30 917/10 in more powerful version and longer wheelbase.
917/K short-tail.
917/L long-tail.
917/PA Spyder for Can-Am series in USA.
920 Porsche: four-/five-speed gearbox for the Type 917.
923 Porsche: 2.0-litre engine for 912E.
924 Volkswagen (Porsche): sports car initially developed for Volkswagen under the number EA 425 and bought back by Porsche.
925 Porsche: four-speed Sportomatic for 911 T/E.
925/09 three-speed Sportomatic for 911.
928 Porsche: sports car with V8 engine, 4.5 litres.
928 S 928 with larger and more powerful engine.
930 Porsche: turbocharged version of the 911.
931 Porsche: supercharged version of 924 (engine 047).
932 Porsche: 924 Turbo with right-hand drive.
934 Porsche: Group 4 racing version of the Type 930.
935 Porsche: Group 5 racing version of the Type 930.
936 Porsche: Group 6 sports car with 2.1-litre turbo engine.
937 Porsche: 924 Carrera GT 937/50: gearbox for 939.
938 Porsche: 924 Carrera GT (right-hand drive).
939 Porsche: 924 Carrera GT 'Le Mans'.
944 Porsche: production sports car with 2.5-litre, four-cylinder engine.
945 Porsche: right-hand drive 944.
956 Porsche: Group C sports car with 2.65-litre turbo engine.
959 Porsche: Group B sports car from 1986 onwards.
960 Experimental car based on the 928.
961 Competitions evolution of 959.
995 Future car for the 1990s (study).
1764 Volkswagen: vehicle and engine.
1778 Volkswagen: 1.3-litre engine.
1817 Volkswagen: bus with diesel engine (study).
1821 Volkswagen: Sportomatic gearbox for Type 1 (Beetle).
1834 Volkswagen: vehicle with 1.3-litre engine.
1837 Volkswagen: 2.5-litre engine and gearbox for Type 1764.
1866 Volkswagen: Beetle successor.
1872 Volkswagen: variant on Type 1866 (study).
1906 Leopard tank.
1966 Volkswagen: small car with four-cylinder underfloor engine (prototypes).
2539 'SAVE' ambulance (research project).
2567 ORBIT fire-fighting system (research project).